ICM-90 Satellite
Conference Proceedings

S. Igari (Ed.)

Harmonic Analysis

Proceedings of a Conference held in Sendai, Japan
August 14-18, 1990

Springer-Verlag
Tokyo Berlin Heidelberg
New York London Paris
Hong Kong Barcelona

Editor
Satoru Igari
Mathematical Institute, Tohoku University
Aobaku, Sendai, Miyagi, 980 Japan

Mathematics Subject Classification (1980): 42-06, 43-06, 43A80, 43A99

ISBN-13:978-4-431-70084-5   e-ISBN-13:978-4-431-68168-7
DOI: 10.1007/978-4-431-68168-7

PREFACE

This volume represents the Proceedings of the Conference on Harmonic Analysis which was held in Sendai, Japan from August 14 to August 18, 1990.

The conference program contained the recent topics and progress of the harmonic analysis on Euclidean spaces and related aspects of real analysis, partial differential equations, complex analysis and probability theory. The studies of these fields have been indirectly inspired by each other and, in some cases, developped by their intimate relations. The intention of the meeting was to participate in an exchange of methods and results in these fields. It included eighteen 50-minute lectures and eighteen short talks.

The conference was financially supported by several sources: The Japan Society for the Promotion of Sciences, Saison Life Insurance Co.Ltd., the Kawai Fundation for Mathematical Sciences, Toyota Motor Co.Ltd. and the Inamori Fundation for young people. The Conference Organizing Committee wishes to express its gratitude to them. We are especially grateful to Saison Life Insurance Co. for its cordial and cooperative support which made this conference possible.

We are also grateful to Tohoku University for all its support and to the very helpful staff of the Mathematical Institute.

Satoru Igari
Sendai, Japan

CONTENTS

Parabolic Harnack Inequalities and Riesz Transforms on Lie Groups of Polynomial growth

G.Alexopoulos

0.Introduction

Let G be a connected Lie group of polynomial growth, i.e. if dg is a left invariant Haar measure and V a compact neighborhood of the identity element e of G , then there are constants $c, d > 0$ such that $dg - measure(V^n) \leq cn^d$, $n \in \mathbb{N}$. Notice that the connected nilpotent Lie groups are of polynomial growth.

Let also $E_1, ..., E_p$ be left invariant vector fields that satisfy the Hörmander condition i.e. they generate together with their successive Lie brackets $[E_{i_1}, [E_{i_2}, ..., E_{i_s}]...]$ the Lie algebra $\mathfrak{g}$ of G. To these vector fields it is associated, in a canonical way, a left invariant distance $d_E(.,.)$ on G, called the controll distance. This distance has the property that (cf.[15]) if $S_E(x, t) = \{y \in G, d_E(x, y) < t\}, x \in G, t > 0$ then there is $c \in \mathbb{N}$ such that

$$S_E(e, n) \subseteq V^{cn}, \quad V^n \subseteq S_E(e, cn), \quad n \in \mathbb{N} \tag{0.1}$$

Moreover the operators $L = (E_1^2 + ... + E_p^2)$ and $\partial/\partial t + L$, according to a classical theorem of Hörmander, are hypoelliptic.

The purpose of this talk is to explain how ideas inspired from Homogeneization theory can be used to answer questions concerning the Harmonic Analysis on G. More precisely, we shall show how a homoneization formula for the operator L can be proved. This formula is similar to the one already known for second order elliptic differetial operators with periodic coefficients on $\mathbb{R}^n$. The novelty here is that we deal with hypoelliptic operators whose coefficients are functions defined on a compact Lie group and not periodic and the homogenised operator L_0 is a left invariant sub-Laplacian (i.e. like L, it is a sum of squares of left invariant vector fields that satisfy the Hörmander condition), defined on a homogeneous nilpotent Lie group N_H and invariant with respect to its dilation structure . N_H is uniquely determined from the algebraic structure of G. Then using a rescaling argument inspired from [2] and [3] and further exploiting the algebraic structure of G, we can obtain the following results.

Theorem 1. *Let* $G, E_1, ..., E_p$ *and* L *be as above. Then for every integer* $k \geq 0, 1 \leq i \leq p$ *and* $0 < a < b < 1$ *there exists* $c > 0$ *such that*

$$|\frac{\partial^k}{\partial t^k} E_i u(at, x)| \leq ct^{-k-\frac{1}{2}} u(bt, x), t \geq 1, x \in G$$

for all $u \geq 0$ *such that* $(\partial/\partial t + L)u = 0$ *in* $(0, t) \times S_E(x, \sqrt{t})$.

Theorem 2. *Let $G, E_1, .., E_p$ and L be as above. Then the operators (Riesz transforms, cf.[12]) $E_i L^{-\frac{1}{2}}, L^{-\frac{1}{2}} E_i, \ 1 \leq i \leq p$, are bounded from L^1 to $weak - L^1$ and from L^q to $L^q, 1 < q < +\infty$.*

If G is also nilpotent then theorem 1 is a particular case of a more general result of Varopoulos [16], namely for all integers $k, \ell \geq 0$ there is $c_{k,\ell} \geq 0$, such that

$$|\frac{\partial^k}{\partial t^k} E_{i_1} ... E_{i_\ell} u(at, x)| \leq ct^{-k-\frac{\ell}{2}} u(bt, x), \ t \geq 1, x \in G \tag{0.2}$$

for all $u \geq 0$ such that $(\partial/\partial t + L)u = 0$, in $(0, t) \times S_E(x, \sqrt{t})$.

These inequalities are also true for $0 < t < 1$ (cf. Varopoulos [15]), but this is a result of the local theory of operators of the type sum of squares of vector fields that satisfy the Hörmander condition.

The motivating example is the universal covering of the group of Euclidean motions on the plane, which is a three dimensional solvable Lie group of polynomial growth . On this group, we can find a sub-Laplacian $L = -(E_1^2 + E_2^2 + E_3^2)$ as above , for which there are families of functions $u_t, v_t, t \geq 1$ and $c > 0$ such that

$$u_t \geq 0, \ Lu_t = 0 \text{ in } S_E(e, t), \ |E_1^2 u_t(e)| \geq \frac{c}{t} u_t(e), \ u_t(e) > c, \ t \geq 1 \tag{0.3}$$

$$v_t \in C_0^\infty(G), \ \|E_1^2 v_t\|_2 \geq ct\|Lv_t\|_2, \ t \geq 1 \tag{0.4}$$

In section 1 we shall discuss this group and we shall show how (0.3),(0.4) and theorems 1 and 2 can be proved. In the subsequent sections we shall indicate how these results can be proved in the general case.

1.The motivating example.

Let Q be a simply connected Lie group of dimension three and assume that there is a basis $\{X_1, X_2, X_3\}$ of its Lie algebra $\mathfrak{q}$ such that

$$[X_1, X_2] = X_3, [X_1, X_3] = -X_2, [X_2, X_3] = 0$$

Identifying the analytical subgroups of Q whose Lie algebra is generated by generated by $\{X_2, X_3\}$ and $\{X_1\}$ with $\mathbb{R}^2$ and $\mathbb{R}$ respectively , we can see that Q is isomorphic to the semidirect product $\mathbb{R}^2 \times_\tau \mathbb{R}$ where the action τ of $\mathbb{R}$ on $\mathbb{R}^2$ is given by $\tau : \mathbb{R} \to L(\mathbb{R}^2)$: $x \to rot_x, rot_x$ being the counterclockwise rotation by angle x and $L(\mathbb{R}^2)$ the space of linear tranformations of $\mathbb{R}^2$.

Q is isomorphic to the universal covering of the group of Euclidean motions on the plane. It is a (non-nilpotent) solvable Lie group of polynomial growth.

Let

$$E_1 = X_1, E_2 = X_1 + X_2, E_3 = X_3 \text{ and } L = -(E_1^2 + E_2^2 + E_3^2)$$

We are going to show how theorems 1 and 2 can be proved in this specific example, construct the families of functions u_t and $v_t, t \geq 1$ mentioned in (0.3) and (0.4) and explain why it is natural to use Homogeneization theory

The fundamental remark is that , if we identify Q with $\mathbb{R}^3$, using the exponential coordinates of the second kind, then L becomes a second order differential operator with periodic coefficients on $\mathbb{R}^3$.

By exponential coordinates of the second kind, we understand the diffeomorphism

$$\phi : \mathbb{R}^3 \to Q, \phi : (x_3, x_2, x_1) \to expx_3 X_3 expx_2 X_2 expx_1 X_1$$

If $x = (x_3, x_2, x_1)$ then we have that

$$d\phi^{-1} E_1(x) = \frac{\partial}{\partial x_1} \quad d\phi^{-1} X_2(x) = \cos x_1 \frac{\partial}{\partial x_2} + \sin x_1 \frac{\partial}{\partial x_3}$$

$$d\phi^{-1} X_3(x) = -\sin x_1 \frac{\partial}{\partial x_2} + \cos x_1 \frac{\partial}{\partial x_3} \tag{1.1}$$

Let us now identify Q with $\mathbb{R}^3$ (as differential manifolds). Then L becomes a uniformly elliptic differential operator , which can be written in divergense form $L = -\frac{\partial}{\partial x_i} a_{ij}(x) \frac{\partial}{\partial x_j}$, with $a_{11} = 2$, $a_{22} = a_{33} = 1$, $a_{12} = a_{21} = \cos x_1$, $a_{13} = a_{31} = \sin x_1$ and $a_{23} = a_{32} = 0$ and the control distanse $d_E(.,.)$ associated to the vector fields E_1, E_2, E_3 equivalent to the Euclidean one i.e. $\exists b \geq a > 0$ such that $a|x - y| \leq d_E(x, y) \leq b|x, y| \; \forall x, y \in \mathbb{R}^3$

Moreover, if $L_\varepsilon = \frac{\partial}{\partial x_i} a_{ij}(\frac{x}{\varepsilon}) \frac{\partial}{\partial x_j}$ and $E_{\varepsilon,i}(x) = E_i(\frac{x}{\varepsilon}), i = 1, 2, 3$, $0 < \varepsilon \leq 1$ and $B(0,1)$ is the Euclidean unit ball then proving the inequalities (0.2) is equivalent to proving that for all $k, \ell \in \mathbb{Z}, k, \ell \geq 0$ and $0 < a < b < 1$ there is $c > 0$ such that

$$|\frac{\partial^k}{\partial t^k} E_{\varepsilon, i_1} ... E_{\varepsilon, i_\ell} u_\varepsilon(a, 0)| \leq cu(b, 0), \; 0 < \varepsilon \leq 1 \tag{1.2}$$

for all $u_\varepsilon \geq 0$ satisfying $(\partial/\partial t + L)u_\varepsilon = 0$ in $(0,1) \times B(0,1)$, which is a problem of Homogeneization theory.

Before we continue we shall recall some results from Homogeneization theory

Results from Homogeneization theory. Let $L = -\frac{\partial}{\partial x_i}a_{ij}(x)\frac{\partial}{\partial x_j}$ be a uniformly elliptic operator in $\mathbb{R}^n$ and assume that its coefficients $a_{ij}(x)$ are periodic (i.e. $a_{ij}(x + z) = a_{ij}(x)$, $x \in \mathbb{R}^n$, $z \in \mathbb{Z}^n$) and Hölder continuous (i.e. there is $\alpha \in (0,1)$ and $M > 0$ such that $\|a_{ij}(x)\|_{C^\alpha(\mathbb{R}^n)} \leq M$).

We denote by $\chi^j, j = 1,...,n$ the unique solutions of the problem

$$L(x_j - \chi^j) = 0, \ \chi^j(x + z) = \chi^j(x), x \in \mathbb{R}^n, z \in \mathbb{Z}^n, \ \int_D \chi^j(x)dx = 0, \ D = [0,1]^n$$

The functions χ^j are called correctors.

We denote by $L_0 = -\frac{\partial}{\partial x_i}q_{ij}\frac{\partial}{\partial x_j}$ the homogenised operator whose coefficients q_{ij} are the constants defined by

$$q_{ij} = \int_D [a_{ij} - a_{i\ell}\frac{\partial}{\partial x_\ell}a_{\ell j}(x)]dx, \ D = [0,1]^n$$

It can be proved that L_0 is an elliptic operator (cf. [4]).

Let $L_\varepsilon = -\frac{\partial}{\partial x_i}a_{ij}(\frac{x}{\varepsilon})\frac{\partial}{\partial x_j}$, $0 < \varepsilon \leq 1$. Let also $f \in L^{n+\delta}(B(0,1))$, $\delta > 0$, $g \in C^{1,\nu}(\partial B(0,1))$, $0 < \nu \leq 1$. and denote by u_ε, $0 \leq \varepsilon \leq 1$, the solutions of the problem

$$L_\varepsilon u_\varepsilon = f \text{ in } B(0,1), \ u_\varepsilon = g \text{ on } \partial B(0,1), \ 0 < \varepsilon \leq 1 \tag{1.3}$$

We have the following results.

Theorem 1.1 (cf. [4]). *Let u_ε, $0 < \varepsilon < 1$ be as above. Then $u_\varepsilon \to 0$, $(\varepsilon \to 0)$, uniformly on the compact subsets of $B(0,1)$*

Theorem 1.2 (cf.[2]). *Let u_ε, $0 \leq \varepsilon \leq 1$. be as above. Then there is a constant $c > 0$ depending only on $\alpha, M, n, \nu, \delta$ and independent of ε such that*

$$[u_\varepsilon]_{C^{0,1}(B(0,1))} \leq c([g]_{C^{1,\nu}(\partial B(0,1))} + \|f\|_{L^{n+\delta}(B(0,1))}) \tag{1.4}$$

In our example, we have that

$$\chi^1(x) = 0, \ \chi^2(x) = \frac{1}{2}\sin x_1, \ \chi^3(x) = -\frac{1}{2}\cos x_1$$

and

$$L_0 = -(2\frac{\partial^2}{\partial x_1^2} + \frac{3}{4}\frac{\partial^2}{\partial x_2^2} + \frac{5}{4}\frac{\partial^2}{\partial x_3^2})$$

L_ϵ can also be written as

$$L_\epsilon = -2\frac{\partial^2}{\partial x_1^2} - 2\cos\frac{x_1}{\epsilon}\frac{\partial^2}{\partial x_1\partial x_2} - 2\sin\frac{x_1}{\epsilon}\frac{\partial^2}{\partial x_1\partial x_3} - \frac{\partial^2}{\partial x_2^2} - \frac{\partial^2}{\partial x_3^2}$$

$$+ \frac{1}{\epsilon}\sin\frac{x_1}{\epsilon}\frac{\partial}{\partial x_2} - \frac{1}{\epsilon}\cos\frac{x_1}{\epsilon}\frac{\partial}{\partial x_3} \quad (1.5)$$

The Harnack inequalities. For $\ell = 1$, (1.2) is a parabolic analogue of (1.4) and it can be proved in a similar way.

Let us now see why (1.2) is not true for $\ell \geq 2$. Let us take $f = 0$ and $g = x_3 + 2$ in (1.3). Then $u_0 = x_3 + 2$. Hence $u_0 \geq 0$, $\frac{\partial}{\partial x_3}u = 1$ and $\frac{\partial}{\partial x_1}u_0 = \frac{\partial}{\partial x_2}u_0 = 0$.

Since $L_\epsilon\frac{\partial}{\partial x_i}u_\epsilon = \frac{\partial}{\partial x_i}L_\epsilon u_\epsilon = 0$, $i = 2,3$, it follows from theorem 1.1 that

$$u_\epsilon \to u_0 \text{ and } \frac{\partial}{\partial x_i}u_\epsilon \to \frac{\partial}{\partial x_i}u_0, \ (\epsilon \to 0), \ i = 2,3 \quad (1.6)$$

uniformly on the compact subsets of $B(0,1)$.

Moreover , it follows from theorem 1.2 that there is $c > 0$ such that

$$\left|\frac{\partial}{\partial x_i}\frac{\partial}{\partial x_j}u_\epsilon(x)\right| \leq c, \ x \in B(0,1) \ \ i = 1,2,3, \ j = 2,3 \quad (1.7)$$

Now, (1.5),(1.6) and (1.7) imply that

$$\frac{\partial^2}{\partial x_1^2}u_\epsilon(0) \sim \frac{1}{\epsilon}, \ (\epsilon \to 0)$$

It follows that the family of functions u_t, $t \geq 1$ defined by $u_t(x) = u_\epsilon(\epsilon x)$, $\epsilon = \frac{1}{t}$ satisfy (0.3).

The Riesz trasforms. The construction of the family v_t, $t \geq 1$, mentioned in (0.4) is similar to the construction of the family u_t, $t \geq 1$ above, i.e. we can consider, in (1.3), $g = 0$ and $f = L_0\varphi$, where $\varphi \in C_0^\infty(B(0,1))$ is such that $\frac{\partial}{\partial x_3}\varphi \neq 0$ and then proceed in the same way.

Let us now see how we can prove that the Riesz transforms $E_i L^{-\frac{1}{2}}$ and $L^{-\frac{1}{2}}E_i$, $i = 1,2,3$ are bounded on L^q, $1 < q < +\infty$ and from L^1 to $weak - L^1$.

It follows from the observation

$$\sum_{1\leq i\leq 3}\|E_i L^{-\frac{1}{2}}\varphi\|_2^2 = -\sum_{1\leq i\leq 3}(E_i^2 L^{-\frac{1}{2}}\varphi, L^{-\frac{1}{2}}\varphi) = (L^{-\frac{1}{2}}\varphi, L^{-\frac{1}{2}}\varphi) = \|\varphi\|_2^2$$

that the tranforms $E_i L^{-\frac{1}{2}}$, as well as their adjoints $L^{-\frac{1}{2}}E_i$, are bounded on L^2. So it is enough to prove that they are bounded from L^1 to $weak - L^1$. Then by interpolation we can prove that they are bounded on L^p, $1 < p < 2$ and by duality on L^p, $2 < p < \infty$

(cf. [11]). Let $K_i(x,y)$ be the kernel of the transform $E_i L^{-\frac{1}{2}}$ and let us use the notation $E_j^y K(x,y)$ to denote the derivation on the variable y with respect to the vector field E_j. If $p_t(x,y)$ is the heat kernel (i.e. the fundamental solution of the equation $(\partial/\partial t + L)u = 0$) then

$$K_i(x,y) = \int_0^\infty t^{-\frac{1}{2}} E_i^x p_t(x,y) dt \text{ and } E_j^y K_i(x,y) = \int_0^\infty t^{-\frac{1}{2}} E_j^y E_i^x p_t(x,y) dt \qquad (1.7)$$

Since the function $u(t,y) = E_i^x p_t(x,y)$ satisfies $(\partial/\partial t + L)u = 0$ theorem 1 can be applied to both $E_i^x p_t(x,y)$ and $E_j^y E_i^x p_t(x,y)$ and using wellknown Gaussian estimates of the heat kernel $p_t(x,y)$ (cf. [1]) we can deduce from (1.7) that there is $c > 0$ such that

$$|K(x,y)| \le \frac{c}{|x-y|^3}, \text{ and } |E_j^y K(x,y)| \le \frac{c}{|x-y|^4} \qquad (1.8)$$

It is wellknown that once we have the estimates (1.8) then it can be proved that $E_i L^{-\frac{1}{2}}$ is bounded from L^1 to $weak - L^1$ (cf.[11]).

Unfortunately, the estimates (1.8) are not availlable for the kernels $K_i^*(x,y)$ of the tranforms $L^{-\frac{1}{2}} E_i$, since in that case we have that

$$K_i^*(x,y) = \int_0^\infty t^{-\frac{1}{2}} E_i^y p_t(x,y) dt \text{ and } E_j^y K_i^*(x,y) = \int_0^\infty t^{-\frac{1}{2}} E_j^y E_i^y p_t(x,y) dt \qquad (1.9)$$

and, as we have seen, the second order Harnack inequalities needed to estimate $E_j^y E_i^y p_t(x,y)$ are not true.

The way to get around this difficulty, is to observe , as it is clear from (1.1), that the natural fields to use are the $\frac{\partial}{\partial x_i}$ and not the E_i, $i = 1,2,3$. Indeed, in that case we can take advantage of the fact that for $i = 2,3$, $\frac{\partial}{\partial x_i}$ and L commute and obtain the adequate estimates for $\frac{\partial}{\partial x_j} \frac{\partial}{\partial x_i} p_t(x,y)$, $i = 2,3$, $j = 1,2,3$ applying theorem 1 twice and then prove that the transforms $L^{-\frac{1}{2}} \frac{\partial}{\partial x_2}$ and $L^{-\frac{1}{2}} \frac{\partial}{\partial x_3}$ are bounded from L^1 to $weak - L^1$ (their L^2 boundedness follows from that of the transforms $L^{-\frac{1}{2}} E_i$, $i = 1,2,3$, proved above).

In order to prove that the transform $L^{-\frac{1}{2}} \frac{\partial}{\partial x_1}$, is bounded from L^1 to $weak - L^1$, we argue as follows. We consider the transform $L^{-\frac{1}{2}} H$ where the vector field H is defined by

$$H' = \frac{\partial}{\partial x_1} + (\frac{\partial}{\partial x_1} x^2) \frac{\partial}{\partial x_2} + (\frac{\partial}{\partial x_1} x^3) \frac{\partial}{\partial x_3} = \frac{\partial}{\partial x_1} + \frac{1}{2} \cos x_1 \frac{\partial}{\partial x_2} + \frac{1}{2} \sin x_1 \frac{\partial}{\partial x_3}$$

Observe that

$$\frac{\partial}{\partial x_1} H = -\frac{1}{2}(L + \frac{\partial^2}{\partial x_2^2} + \frac{\partial^2}{\partial x_3^2} + \cos x_1 \frac{\partial^2}{\partial x_1 \partial x_2} + \sin x_1 \frac{\partial^2}{\partial x_1 \partial x_3})$$

So, as it has already been explained above, we can estimate $\frac{\partial}{\partial x_i} H p_t(x,y)$, $i = 1,2,3$ applying theorem 1 twice and prove the estimates (1.8) for the kernel $K_H(x,y)$ of the transform $L^{-\frac{1}{2}} H$, which in turn implies that $L^{-\frac{1}{2}} H$, hence $L^{-\frac{1}{2}} \frac{\partial}{\partial x_1}$, is bounded from L^1 to $weak - L^1$. In the following sections, we shall try to explain how these ideas can be generalised on any connected Lie group of polynomial growth.

2. The structure of the Lie algebra.

The connected Lie groups of polynomial growth have been characterised by Y.Guivarc'h [8] as those whose Lie algebra $\mathfrak{g}$ is of type R, i.e. all the eigenvalues of the derivations $adX(Y) = [X, Y]$ are imaginary.

So, let us fix an R-type Lie algebra $\mathfrak{g}$ and denote by $\mathfrak{q}, \mathfrak{n}$ and $\mathfrak{m}$ respectively the radical, the nil-radical and a Levi sub-algebra of $\mathfrak{g}$. $\mathfrak{q}$ and $\mathfrak{n}$ are, respectively, solvable and nilpotent ideals and $\mathfrak{m}$ a semisimple subalgebra of $\mathfrak{g}$. Moreover (cf.[13])

$$\mathfrak{n} \subseteq \mathfrak{q}, \quad \mathfrak{g} = \mathfrak{q} + \mathfrak{m}, \quad \mathfrak{q} \cap \mathfrak{m} = \{0\}, \quad [\mathfrak{q}, \mathfrak{g}] \subseteq \mathfrak{n} \tag{2.1}$$

We denote by $adX = S(X) + K(X)$ the Jordan decomposition of the derivation $adX(Y) = [X, Y]$, $X \in \mathfrak{g}$. $S(X)$ is the semisimple and $K(Y)$ the nilpotent part. It is wellknown that

(i) $S(X)$ and $K(X)$ are derivations of $\mathfrak{g}$ (cf.[13])

(ii) $[S(X), S(Y)] = 0$, $X, Y \in \mathfrak{q}$ (cf.[15], appendix 1.6) $\tag{2.2}$

Let $\mathfrak{q}_0 = \{X \in \mathfrak{q} : [Z, X] = 0, Z \in \mathfrak{m}\}$. Clearly $\mathfrak{q}_0$ is a subalgebra of Q. It follows from (2.1) and the semisimplicity of $\mathfrak{m}$ that there is a linear subspace $\mathfrak{q}_1 \subseteq \mathfrak{q}$ such that

$$\mathfrak{q} = \mathfrak{q}_0 \oplus \mathfrak{q}_1, \quad \mathfrak{q}_1 \subseteq \mathfrak{n} \text{ and } adX(\mathfrak{q}_1) \subseteq \mathfrak{q}_1, \quad X \in \mathfrak{m}$$

Since $\mathfrak{q}_0$ is a solvable algebra itself, using (2.2) we can find and fix once and for all, vector fields $X_1, ..., X_k \in \mathfrak{q}_0$, $k = dim\mathfrak{q}/\mathfrak{n}$, such that

a) $S(X)X_i = 0$, $adZ(X_i) = 0$, $X \in \mathfrak{q}$, $Z \in \mathfrak{m}$, $i = 1, ..., k$.

b) The images of $X_1, ..., X_k$ by the map $\mathfrak{q} \to \mathfrak{q}/\mathfrak{n}$ form a basis $\mathfrak{q}/\mathfrak{n}$

The filtration of $\mathfrak{q}$. We define a filtration $\mathfrak{q} = \mathfrak{r}_1 \supseteq ... \supseteq \mathfrak{r}_m \supseteq \mathfrak{r}_{m+1} = 0$, $\mathfrak{r}_m \neq \{0\}$ of $\mathfrak{q}$, in the following way :

1) We put $\mathfrak{r}_1 = \mathfrak{q}$.

2) We define $\mathfrak{r}_2$ as the linear subspace of $\mathfrak{q}$ generated by the vectors $[X_i, X_j], 1 \leq i, j \leq k$ and the spaces $K(X)\mathfrak{n}$, $X \in \mathfrak{q}$ and

3) if $i \geq 2$ and $\mathfrak{r}_i$ has already been defined, the $\mathfrak{r}_{i+1}$ is defined as the linear subspace of $\mathfrak{q}$ generated by the spaces $K(X)\mathfrak{r}_i$, $X \in \mathfrak{q}$

The most important properties of this filtration are gathered in the next proposition.

Proposition 1.1. *1)* $\mathfrak{r}_i$ *is an ideal of* $\mathfrak{q}$, $i = 1, 2, ...$
2) If $\mathfrak{r}_i \neq \{0\}$, *then* $\mathfrak{r}_i \neq \mathfrak{r}_{i+1}$. *Therefore there is* $m \in \mathbb{N}$ *such that* $\mathfrak{r}_m \neq \{0\}, \mathfrak{r}_{m+1} = \{0\}$.
3) $[\mathfrak{r}_i, \mathfrak{r}_j] \subseteq \mathfrak{r}_{i+j}$, $i, j \geq 2$.
4) There are subspaces $\mathfrak{a}_1, ..., \mathfrak{a}_m$ *of* $\mathfrak{q}$ *such that*

a) $adZ(\mathfrak{a}_i) \subseteq \mathfrak{a}_i$, $S(X)\mathfrak{a}_i \subseteq \mathfrak{a}_i$, $Z \in \mathfrak{m}$, $X \in \mathfrak{q}$, $i = 1, ..., m$ *and*

b) $\mathfrak{r}_i = \mathfrak{a}_i \oplus ... \oplus \mathfrak{a}_m$

We $n = dim\mathfrak{q}/\mathfrak{n}$, $n_0 = 0$ and $n_i = dim(\mathfrak{a}_1 \oplus ... \oplus \mathfrak{a}_i)$, $i = 1, ..., m$. Then

$$1 \leq k \leq n_1 < ... < n_m = n$$

The nil-shadow $\mathfrak{q}_N$ *of* $\mathfrak{q}$. We can verify directly that the conditions

$$[X_i, X_j]_N = [X_i, X_j], \; [X_i, Y]_N = K(X_i)Y, \; [Y, Z]_N = [Y, Z], \; 1 \le i, j \le k, \; Y, Z \in \mathfrak{n}$$

define a Lie product on the linear space $\mathfrak{q}$. The Lie algebra $\mathfrak{q}_N = (\mathfrak{q}, [\, , \,]_N)$ is nilpotent and it is called the nil-shadow of $\mathfrak{q}$.

The choice of the basis of $\mathfrak{q}$. It follows from (2.2) that the linear transformations $S(X)$, $X \in \mathfrak{q}$ commute. Moreover, since $\mathfrak{q}$ is of type R, all their eigenvalues are purely imaginary, i.e. of the type ia, $a \in \mathbb{R}$. Using these two observations we can construct a basis $\{X_1, ..., X_n\}$ of $\mathfrak{q}$ which will be fixed once and for all, such that

    1) $X_1, ..., X_k$ are as above.

    2) $\{X_{n_i+1}, ..., X_{n_{i+1}}\}$ is a basis of $\mathfrak{a}_{i+1}$, $i = 1, ..., m$ and

    3) for all $i = 1, ..., n$, either $e^{S(X)}X_i = X_i$, $\forall X \in \mathfrak{q}$, or there is a linear form $\lambda_i : \mathfrak{q} \to \mathbb{R}$ and $j = i$ or $i + 1$ such that, $\forall X \in \mathfrak{q}$, we have

$$\begin{aligned}
e^{S(X)}X_i &= \cos \lambda_i(X)X_i + \sin \lambda_i(X)X_j \\
e^{S(X)}X_j &= -\sin \lambda_i(X)X_i + \cos \lambda_i(X)X_j
\end{aligned} \tag{2.3}$$

3. The exponential coordinates of the second kind.

Let G be a connected Lie group of polynomial growth and $\mathfrak{g}$ its Lie algebra. We identify the elments of $\mathfrak{g}$ with the left invariant vector fields on G.

Let $\mathfrak{q}$, $\mathfrak{n}$ and $\mathfrak{m}$ be as in section 2 and denote by Q, N and M the analytical subgroups of G having Lie algebras $\mathfrak{q}$, $\mathfrak{n}$ and $\mathfrak{m}$ respectively. $S(X)$, $K(X)$, $X \in \mathfrak{g}$ and the basis $\{X_1, ..., X_n\}$ of $\mathfrak{q}$ are as in section section 2.

We have that $G = QM$. Moreover, if we denote by φ the diffeomorphism

$$\varphi : Q \times M \to G, \; \varphi : (x, z) \to xz$$

and by Adz the differential of the inner automorphim $g \to z^{-1}gz$ of G , then

$$d\varphi^{-1}(X + Z)(x, z) = (Adz(X)(x), Z(z)), \; X \in \mathfrak{q}, Z \in \mathfrak{m}, x \in Q, z \in M \tag{3.1}$$

The simply connected case. Let $\bar{G}$ be the universal covering of G. Then $\bar{G}$ is simply connected and we denote by $\bar{Q}, \bar{N}$ and $\bar{M}$ the analytical subgroups of $\bar{G}$ whose Lie algebra is $\mathfrak{q}, \mathfrak{n}$ and $\mathfrak{m}$ respectively.

It is wellknown (cf. [13]) that the map

$$\phi : \mathbb{R}^n \to \bar{Q}, \; \phi : x = (x_n, ..., x_1) \to \exp x_n X_n ... \exp x_1 X_1$$

is a diffeomorphism, which is called exponential coordinates of the second kind. We want to give an expresion for $d\phi^{-1}$. To this end, we shall need some notations.

We denote by $\bar{K}(X_i)$, the linear transformation of $\mathfrak{q}$ defined by

$$\bar{K}(X_i)X_j = 0, \text{ for } i \geq j \text{ and } \bar{K}(X_i)X_j = K(X_i)X_j, \text{ for } i < j$$

If $B(x) = b_n(x)\frac{\partial}{\partial x_n} + ... + b_1(x)\frac{\partial}{\partial x_1}$ is a vector field on $\mathbb{R}^n$, then we put $pr_iB(x) = b_i(x)$. We also use the same notation for left invariant vector fields on Q, i.e. if $E = c_nX_n + ... + c_1X_1$, then we put $pr_iE = c_i$.

We have the following formula :

$$pr_i d\phi^{-1}E(x) = pr_i e^{x_n \bar{K}(X_n)}...e^{x_1 \bar{K}(X_1)}e^{x_k S(X_k)}...e^{x_1 S(X_1)}E \qquad (3.2)$$

Let $\bar{Q}_N$ be a simply connected nilpotent Lie group that admits as Lie algebra the nil-shadow $\mathfrak{q}_N$ of $\mathfrak{q}$. $\bar{Q}_N$ is also called the nil-shadow of $\bar{Q}$. Using again the exponential coordinates of the second kind we can see that $\bar{Q}_N$ is diffeomorphic with $\mathbb{R}^n$.

From now on, we shall identify $\bar{Q}$ and $\bar{Q}_N$ as differential manifolds with $\mathbb{R}^n$.

We identify the elements of $\mathfrak{q}_N$ with the left invariant vector fields on Q_N and if $X \in \mathfrak{q}$ then we denote by $_NX$ the element of $\mathfrak{q}_N$ satisfying $_NX(0) = X(0)$. We extend the transformations $S(X)$, $X \in \mathfrak{q}$ and Adz, $z \in M$ to $\mathfrak{q}_N$ by putting $S(X)_NY = _N(S(X)Y)$ and $Adz(_NY) = _N(Adz(Y))$.

It follows from (3.2) that if $x = (x_n, ..., x_1) \in \mathbb{R}^n$ and $E \in \mathfrak{q}$ then

$$E(x) = e^{x_k S(X_k)}...e^{x_1 S(X_1)}{}_N E(x) \qquad (3.3)$$

Using the diffeomorphism

$$\Phi : \mathbb{R}^n \times \bar{M} \to \bar{G} = \bar{Q}\bar{M}, \ \Phi : (x,z) \to \phi(x)z$$

we identify the groups $\bar{G}$ and $\bar{Q}_N \times \bar{M}$ as differential manifolds with $\mathbb{R}^n \times \bar{M}$. Also, if $E = (X, Z)$ is a vector field on $\mathbb{R}^n \times \bar{M}$ then we write $E = X + Z$.

Putting (3.1) and (3.3) together we have that

$$d\Phi^{-1}(X + Z)(x,z) = e^{x_k S(X_k)}...e^{x_1 S(X_1)}Adz(_NX)(x,z) + Z(x,z),$$
$$X \in \mathfrak{q}, \ Z \in \mathfrak{m}, \ x = (x_n, ..., x_1) \in \mathbb{R}^n, z \in \bar{M} \quad (3.4)$$

The fundamental group of G. Let Γ be the fundamental group of G. Then Γ is a finitely generated abelian group. Therefore it can be writen as $\Gamma = \Gamma_1 \times A$ where A is a finite abelian subgroup and Γ_1 is a subgroup isomorphic with $\mathbb{Z}^d$ for some $d \leq n$.

We can easily see that $A \subseteq \bar{M}$ and $\Gamma_1 \subseteq \bar{Q}$. So $M = \bar{M}/A$ and $Q = \bar{Q}/\Gamma_1$.

It can be proved that the the basis $\{X_n, ..., X_1\}$ of $\mathfrak{q}$ can be chosen in such a way that

$$expx_{i_1}X_{i_1}, ..., expx_{i_d}X_{i_d} \text{ generate } \Gamma_1 \text{ for some integers } 1 \leq i_1 < ... < i_d \leq n$$

From now on, we shall assume that the basis $\{X_n, ..., X_1\}$ (chosen in section 2 and fixed since then) has this additional property.

Γ_1 is also a subset of $\bar{Q}_N$, since both $\bar{Q}_N$ and $\bar{Q}$ have been identified as differential manifolds with $\mathbb{R}^n$. It can also be proved that Γ_1 is a subgroup of $\bar{Q}_N$. The group $Q_N = \bar{Q}_N/\Gamma_1$ is called the nil-shadow of Q.

The non-simply connected case. We define $\mathbb{O}_i = \mathbb{T}(= \mathbb{R}/\mathbb{Z})$, if $expX_i \in \Gamma_1$ and $\mathbb{O}_i = \mathbb{R}$, if not, $i = 1, ..., n$ and we put $\mathbb{O} = \mathbb{O}_n \times ... \times \mathbb{O}_1$.

As in the simpy connected case, we have the diffeomorphisms, which we shall also denote by ϕ and Φ :

$$\phi : \mathbb{O} \to Q, \ \phi : x = (x_n, ..., x_1) \to expx_n X_n ... expx_1 X_1$$

$$\Phi : \mathbb{O} \times M \to G = QM, \ \Phi : (x, z) \to \phi(x)z$$

Similarly, using ϕ and Φ respectively, we identify the groups Q, Q_N with $\mathbb{O}$ and $G, Q_N \times M$ with $\mathbb{O} \times M$ as differential manifolds. If $X \in \mathfrak{q}$ is a left invariant vector field on Q, then we denote by $_N X \in \mathfrak{q}_N$ the left invariant vector field on Q_N that satisfies $_N X(0) = X(0)$. If $E = (X, Z)$ is a vector field on $\mathbb{O} \times M$, then we write $E = X + Z$. With these changes in the notations (3.4) remains true, i.e.

$$d\Phi^{-1}(X + Z)(x, z) = e^{x_k S(X_k)} ... e^{x_1 S(X_1)} Adz(_N X)(x, z) + Z(x, z),$$
$$X \in \mathfrak{q}, \ Z \in \mathfrak{m}, \ x = (x_n, ..., x_1) \in \mathbb{O}, z \in \bar{M} \quad (3.5)$$

4. The volume growth.

Let G be a connected Lie group of polynomial growth, dg a left invariant Haar measure on G and $\mathbb{O} \times M$ as in section 3. As it was explained in section 3 we identify G and $\mathbb{O} \times M$ as differential manifolds. If e is the identity element of M, then we denote by 0 the identity element $(0, e)$ of G.

Let $n_0, n_1, ..., n_m$ be as in section 2 and put

$$\sigma(i) = 0, \text{ if } \mathbb{O}_i = \mathbb{T} \text{ and } \sigma(i) = j, \text{ if } n_{j-1} < i \le n_j, \ \mathbb{O}_i = \mathbb{R}, \ i = 1, ..., n.$$

$$d = \sigma(1) + ... + \sigma(n) \quad (4.1)$$

Let $E_1, ..., E_p$ be as in theorem 1, i.e. left invariant vector fields on G that satisfy the Hörmander condition. The control distanse $d_E(., .)$ associated to these vector fields is defined as follows :

We call, a piecewise smooth path $\dot{\gamma} : [0, 1] \to G$ admissible if and only if $\dot{\gamma}(t) = a_1(t)E_1 + ... a_p(t)E_p$ and we put $|\dot{\gamma}(t)|^2 = a_1^2(t) + ... + a_p^2(t)$. Then we define

$$d_E(x, y) = \inf\{\int_0^1 |\dot{\gamma}(t)|^2 dt, \ \gamma \text{ admissible path such that } \gamma(0) = x, \ \gamma(1) = y\}$$

We put $\quad S_E(x, t) = \{y \in G : d_E(x, y) < t\}, \ x \in G, \ t > 0.$

The goal of this section is to give an expression for $dg - measure(S_E(0,t)$ and to study the shape of the balls $S_E(0,t)$, $t \geq 0$. As we see from (0.1) they behave for large t in the same way as the powers V_n, $n \in \mathbb{N}$ of a compact neighborhood V of 0. Hence the vector fields $\{E_1, ..., E_p\}$ can be replaced with a basis $\{X_n, ..., X_1, Z_1, ..., Z_r\}$ of the Lie algebra $\mathfrak{g}$ of G, $Z_1, ..., Z_r \in \mathfrak{m}$. The next thing to observe is that, it follows from (3.5), that $\{X_n, ..., X_1, Z_1, ..., Z_r\}$ can be replaced by $\{_N X_n, ..., _N X_1, Z_1, ..., Z_r\}$. In this way questions concerning the volume growth of G are reduced to questions cocerning the volume growth of $Q_N \times M$, where the answers are wellknown (cf. [7],[8],[16]). So we have the following

Proposition 4.1. Let $S_E(x,t)$, d be as above and put $D_t = \{((x_n, ..., x_1), z) \in \mathbb{O} \times M$ such that, $x_i < t^{\sigma(i)}\}$, $t \geq 1$. Then there is $c > 0$ such that

$$S_E(0, c^{-1}t) \subseteq D_t \subseteq S_E(0, ct), \ t \geq 1$$
$$c^{-1}t^d \leq dg - measure(S_E(0,t)) \leq ct^d, \ t \geq 1$$

5. The Homogeneization formula.

In this section, we shall indicate how some classical results of Homogeneization theory (cf. [4]) can be generalised in our case .

Let G be a connected Lie group of polynomial growth which, as it was explained in section 3, we identify, as a differential manifold, with $\mathbb{O} \times M$. Let $n_0, n_1, ..., n_m$ be as in section 2 and $\sigma(i)$, $i = 1, ..., n$ as in section 4. Let $E_1, ..., E_p$ and L be as in theorem 1, i.e. $F_1, ..., F_p$ are left invariant vector fields on G that satisfy the Hörmander condition and $L = -(E_1^2 + ... + E_p^2)$. The purpose of this section is to indicate how we can prove a Homogeneization formula for the operator L.

The dilation. The dilation τ_ε, $0 < \varepsilon \leq 1$ of $\mathbb{O} \times M$, is defined to be the transformation

$$\tau_\varepsilon : \mathbb{O} \times M \to \mathbb{O} \times M, \ \tau_\varepsilon : ((x_n, ..., x_1), z) \to ((\varepsilon^{\sigma(n)} x_n, ..., \varepsilon^{\sigma(1)} x_1), z)$$

We put

$$E_{\varepsilon,i} = \frac{1}{\varepsilon} d\tau_\varepsilon(E_i), \ i = 1, ..., p \text{ and } L_\varepsilon = -(E_{\varepsilon,1}^2 + ... + E_{\varepsilon,p}^2), \ 0 < \varepsilon \leq 1$$

The compactness. If $(s, x) \in \mathbb{R} \times G$ and $u \in C^\infty([s - \rho^2, s]) \times S_E(x, \rho))$, then we write

$$Osc(u, s, x, \rho) = sup\{|(t, y) - u(t', y')|, \ (t, y), (t', y') \in [s - \rho^2, s] \times S_E(x, \rho)\}$$

We also put $D = \{((x_n, ..., x_1), z) \in \mathbb{O} \times M : |x_i| < 1 \ i = 1, ..., n, \ z \in M\}$

Theorem 5.1 (cf.[10]). *For every $0 < \delta < 1$, there is $0 < a < 0$ such that*

$$Osc(u, s, x, \delta\rho) \leq aOsc(u, s, x, \rho), \ (s, x) \in \mathbb{R} \times G$$

for all $u \in C^\infty([s-\rho^2, s]) \times S_E(x, \rho))$ such that $(\frac{\partial}{\partial t}+L)u = 0$ in $[s-\rho^2, s]) \times S_E(x, \rho)$, $\rho > 0$.

The above theorem provides a compactness on the space of functions u_ε, satisfying

$$\|u_\varepsilon\|_\infty \leq 1, \ (\partial/\partial t + L_\varepsilon)u_\varepsilon = 0 \text{ in } (-1, 1) \times D, \ 0 < \varepsilon \leq 1 \tag{5.1}$$

In particular we have the following

Proposition 5.2. *Let $u_\varepsilon, 0 < \varepsilon \leq 1$ be a family of functions satisfying (5.1). Then there is a subsequense, also denoted by u_ε, such that*

$$u_\varepsilon \to u_0, \ (\varepsilon \to 0)$$

uniformly on the compact subscts of $(-1, 1) \times D$.
Moreover, $u_0(t, ((x_n, ..., x_1), z)) = u_0(t, ((x'_n, ..., x'_1), z'))$, for all
$((x_n, ..., x_1), z), \ ((x'_n, ..., x'_1), z') \in D$, *such that $x_i = x'_i$ if $\mathbb{O}_i = \mathbb{R}$, $i = 1, ..., n$.*
     Let $\mathbb{O}_c = \{x = (x_n, ..., x_1) \in \mathbb{O} : x_i = 0$ if $\mathbb{O}_i = \mathbb{R}$, $i = 1, ..., n\}$, $\mathbb{O}_H = \mathbb{O} \times M/\mathbb{O}_c \times M$ and D_H the image of D by the natural map $\mathbb{O} \times M \to \mathbb{O}_H$. Then we have the following

Corollary 5.3. *The limit function u of the proposition 5.2 can be viewed as defined on D_H.*

The limit group N_H. Let $X_n, ..., X_1$ be the basis of the radical $\mathfrak{q}$ of the Lie algebra $\mathfrak{g}$ of G used to construct $\mathbb{O} \times M$ in section 3. Let $K(X)$ and $\mathfrak{a}_1, ..., \mathfrak{a}_m$ be as in section 2. Then $\mathfrak{q} = \mathfrak{a}_1 \oplus ... \oplus \mathfrak{a}_m$. We denote by $pr_{\mathfrak{a}_i}$ the projection on $\mathfrak{a}_i$. We denote by $\mathfrak{q}_H = (\mathfrak{q}, [., .]_H)$ the Lie algebra defined on the space $\mathfrak{q}$ by the condition

$$[X_i, X_j]_H = pr_{\mathfrak{a}_{\sigma(i)+\sigma(j)}} K(X_i)X_j, \ i \leq j$$

Clearly, $\mathfrak{q}_H$ is a nilpotent Lie algebra of homogeneous type Let $\mathfrak{b}$ be the sub- algebra of $\mathfrak{q}_H$ generated by the vectors $\{X_i : \mathbb{O}_i = \mathbb{T}, \ i = 1, ..., n\}$. Let Q_H be a simply connected Lie group that admits $\mathfrak{q}_H$ as its Lie algebra and denote by C its analytical subgroup of Q_H having $\mathfrak{b}$ as Lie algebra.
     The limit group N_H is defined to be the quotient $N_H = Q_H/C$. It is a simply connected nilpotent Lie group of homogeneous type.
     Observe that if we identify N_H, as a differential manifold, with $\mathbb{O}_H$ (using the exponential coordinates of the second kind) then the limit function u_0 in the proposition 5.2 becomes a function defined on N_H.
     The correctors. The correctors $\chi^j(x, z)$, $j = 1, ..., n_1$, $((x, z) = ((x_n, ..., x_1), z) \in \mathbb{O} \times M)$, are defined to be functions periodic with respect to the variable x, wth mean 0, satisfying $L(x_j - \chi^j) = 0$. We can see that they depend only on $x_k, ..., x_1$ and z (k was defined, in section 1, to be $k = dim\mathfrak{q}/\mathfrak{n}$.
     The homogenised operator L_0. For $1 \leq i, j \leq n_1$, let q_{ij} denote the mean of the function $(E_1 x_i)E_1(x_j - \chi^j) + ... + (E_p x_i)E_p(x_j - \chi^j)$ (observe that it is quasiperiodic with respect to the variable x). Let also $L'_0 = -\sum_{1 \leq i, j \leq n_1} \frac{\partial}{\partial x_i} q_{ij} \frac{\partial}{\partial x_j}$. It can be shown that L'_0 is an

elliptic operator in $\mathbb{R}^{n_1}$. Therefore there are linearly independent constant vector fields $Y_1, ..., Y_{n_1}$ such that $L_0' = -(Y_1^2 + ... + Y_{n_1}^2)$. Using the linear space isomorphism defined by $\frac{\partial}{\partial x_i} \to X_i$, $i = 1, ..., n_1$ to identify the Lie algebra of $\mathbb{R}^{n_1}$ with $\mathfrak{a}_1$, we can view $Y_1, ..., Y_{n_1}$ as elements (a basis actually) of $\mathfrak{a}_1$, hence of $\mathfrak{q}_H$. Identifying in turn the elements of $\mathfrak{q}_H$ with the left invariant vector fields on Q_H, L_0' becomes an operator on Q_H which we denote by L_H.

The homogenised operator L_0 is defined to be the image of L_H by the natural map $Q_H \to N_H = Q_H/C$.

Now we can state the following

Proposition 5.4. *The limit function u_0 can be viewed as defined on N_H and then it satisfies $(\partial/\partial t + L_0)u_0 = 0$ in $(-1, 1) \times D_H$.*

6. The proof of theorems 1 and 2.

Let us fix a basis $\{Z_1, ..., Z_r\}$ of the Lie algebra $\mathfrak{m}$ of M. We observe that, with the notations of section 5, theorem 1 is equivalent to proving that there is $c > 0$, independent of ε, such that

$$|\frac{\partial}{\partial t} u_\varepsilon(0,0)| \leq c, \quad |\frac{\partial}{\partial x_i} u_\varepsilon(0,0)| \leq c, \ 1 \leq i \leq n_1$$

$$|Z_i u_\varepsilon(0,0)| \leq c, \ 1 \leq i \leq r, \quad |\frac{\partial}{\partial x_i} u_\varepsilon(0,0)| \leq c\varepsilon^{-\sigma(i)+1}, \ n_1 < i \leq n$$

for all u_ε satisfying $\|u_\varepsilon\|_\infty \leq 1$ and $(\partial/\partial t + L_\varepsilon)u_\varepsilon = 0$ in $(-1, 1) \times D$, $0 \leq \varepsilon \leq 1$.

These inequalities can be proved by adapting the proof of theorem 2 of [2] (also cf.[3]). It should be mentioned that the correctors χ^j, $1 \leq j \leq n_1$, as well as proposition 5.4, are used in a fundamental way.

The strategy in proving theorem 2 is the following. As in section 1 we observe that the transforms presenting difficulties are the $L^{-\frac{1}{2}} E_i$, $i = 1, ..., p$. We split their kernels $K_i(x, y) = \int_0^\infty t^{-\frac{1}{2}} E_i p_t(x, y) dt$, $(p_t(x, y)$ is the heat kernel), in two parts : $K_1(x, y) = \int_0^1 t^{-\frac{1}{2}} E_i p_t(x, y) dt$ and $K_2(x, y) = \int_1^\infty t^{-\frac{1}{2}} E_i p_t(x, y) dt$. Using the Gaussian estimates for $p_t(x, y)$ proved in [15] and the inequalities (0.2) we can prove the appropriate estimates for the kernel $K_1(x, y)$ and then deduce that the corresponding operator $K_1 f(x) = \int K_1(x, y) f(y) dy$ is bounded on L^q, $1 < q < \infty$ and from L^1 to $weak - L^1$.

To handle the kernel $K_2(x, y)$ we consinder the kernels $R_i(x, y) = \int_1^\infty t^{-\frac{1}{2}} H_i p_t(x, y) dt$, $-r \leq i \leq n$, $i \neq 0$, where $H_i = Z_{-i}$, for $-r \leq i < 0$, $H_i = _N X_i$, for $k < i \leq n$ and $H_i = _N X_i + \sum_{k<i\leq n_1} (_N X_i \chi^j)_N X_j$, for $1 \leq i \leq k$. It turns out (for reasons that become clear from the proof of theorem 1, also cf. section 1) that we can control the derivatives $E_i H_j p_t(x, y)$, $t \geq 1$ and therefore obtain the appropriate estimates for the kernels $R_j(x, y)$ which in turn allow us to prove that the corresponding operators, hence the operator $K_2 f(x) = \int K_2(x, y) f(y) dy$, are bounded on L^q, $1 < q < \infty$ and from L^1 to $weak - L^1$. We should mention that to prove the L^2 boundedness we use the $T1$ theorem.

Once the L^1 to $weak - L^1$ boundedness of the transforms $L^{-\frac{1}{2}}E_i$, $i = 1,...,p$ is established, the proof of theorem 2 can be concluded by standard arguments (cf. section 1).

Bibliography

[1] D.C.Aronson, Bounds for the fundamental solution of a parabolic equation, *Bull. Amer. Math. Soc.* **73**, (1967), 890-896.

[2] M.Avelaneda and F.H.Lin, Compactness methods in the theory of Homogeneization, *Comm. Pure Appl. Math.* **40** (1987), 803-847.

[3] M.Avelaneda and F.H.Lin, Un théorème de Liouvile pour des équations elliptiques avec coefficients périodiques, *C.R.Acad.Paris*, **t. 609**, serie I, p. 245-250, 1989.

[4] A.Bensoussan, J.L. Lions andG.Papanicolaou, *Asymptotic analysis of periodic structures* , North Holland Publ., 1978.

[5] R.Coifman and G.Weis, *Lecture Notes in Mathematices* Sringer-Verlag, 1984

[6] G.David and J.L.Journé, A boundedness criterion for generalised Calderon-Zygmund operators, *Annals of Math.*, **120**, no12, (1984), 371-378.

[7] G.B.Folland and E.Stein, *Hardy spaces on Homogeneous groups*, Princeton University Press, 1982.

[8] Y.Guivarch, Croissance polynômiale et périodes de fonctions harmoniques, *Bull. Sc. Math. France*, **101** (1973), 149-152.

[9] L.Hörmander, Hypoelliptic second order operators, *Acta Math.* **119** (1967), 147-171.

[10] L.Saloff-Coste, Analyse sur les groupes de Lie à croissance polynômiale, *C.R.Acad.Paris*, **309**, serie I, 1989, p.149-152.

[11] E.Stein, *Singular integrals and differentiability properties of functions*, Ann. of Math. Studies (1976)

[12] E.Stein, *Topics in Harmonic Analysis*, Princeton University Press, 1970.

[13] V.S.Varadarajan, *Lie groups, Lie algebras and their representations*, Springer-Verlag, 1984.

[14] N.Th.Varopoulos, Fonctions harmoniques positives sur les groupes de Lie, *C.R.Acad. Paris*, **309**, serie I, 1987, p.519-521.

[15] N.Th.Varopoulos, Analysis on Lie groups, *J. Funct. Anal.*, **76**, no 2, 1988, p.346-410.

[16] N.Th.Varopoulos, *Lecture Notes*, Université de Paris VI, Notes taken by L.Saloff-Coste and T.Coulhon.

Professor George ALEXOPOULOS

Department of Mathematics
McGill University
805 Sherbrooke St. West
Montreal, QC, CANADA H3A 2K6

Harmonic Analysis with respect to Degenerate Laplacians on Strictly Pseudoconvex Domains

Hitoshi Arai

Mathematical Institute, Tohoku University

Aobaku, Sendai 980, JAPAN

In this paper we will study the harmonic analysis associated with the Laplace-Beltrami operators of the Bergman metrics or other related complete Kähler metrics of strictly pseudoconvex domains. Moreover, for this purpose, we will investigate potential theory for a broad class of open Riemannian manifolds which contains strictly pseudoconvex domains. We will deal with the following subjects.

1. Boundary Harnack Principle.
2. Minimal Martin Boundary.
3. Dirichlet Problem.
4. Regularity of Degenerate elliptic Harmonic Measures.
5. Boundary Behavior of Degenerate Harmonic Functions.

In section 1 we describe our notation and introduce a class of Riemannian manifolds which includes strictly pseudoconvex domains with their Bergman metrics and, in addition, Cartan-Hadamard manifolds with bounded negative sectional curvatures. Section 2 is devoted to the potential theory on the Riemannian manifolds, namely topics 1,2 and 3 mentioned above are discussed. Section 3 and 4 deal with the applications to potential theory on strictly pseudoconvex domains endowed with their Bergman metrics or other related Kähler metrics. In section 5 and 6 we study subjects 4 and 5.

§1. **Riemannian manifolds modeled on strictly pseudoconvex domains.** Let $D \subset \mathbf{C}^n$ be a bounded strictly pseudoconvex domain with C^∞ boundary ∂D. Denote by g_b the Bergman metric of D. By a result of C. Fefferman [4] we have known that the main part of the metric is the following one: Let λ be a smooth strictly plurisubharmonic function defined on a neighborhood U of the closure $\overline{D}$ of D such that

$$| \nabla\lambda(\zeta) |> 0 \text{ for } \zeta \in \partial D \text{ and } D = \{z \in U : \lambda(z) < 0\}$$

From this function one can make the metric

$$g = (g_{i\bar{j}}) \; ; \; g_{i\bar{j}}(z) = -\frac{\partial^2}{\partial z_i \partial \bar{z}_j} \log(-\lambda(z)), \; z \in D$$

It is well known that g is a complete Kähler metric of D. The Fefferman's result tell us that the Bergman metric is asymptotically quasi-isometric to the metric g.

In order to study systematically the Kähler manifolds (D, g_b) and (D, g), we introduce a class of Riemannian manifolds:

Definition 1 *An open N-dimensional Riemannian manifold (M, h) is said to satisfy condition (α) if it posseses the following two conditions (A) and (B):*

(A) (Bounded Geometry Property) There exists $r_0 > 0$ such that for every $x \in M$ there is a coordinate map χ_x from the geodesic ball $B(x, r_0)$ with center x and radius r_0 to $\mathbf{R}^N$ satisfying

$$\frac{1}{c_0} d(y, z) \leq | \chi_x(y) - \chi_x(z) | \leq c_0 d(y, z)$$

for all $y, z \in B(x, r_0)$, where c_0 is a positive constant independent of x, y, and z, and $d(,)$ is the Riemannian distance function of (M, h).

(B) (M, h) has its compactification $\hat{M}$ and an open set Ω in M satisfying

(i) The intersection $\partial_\Omega M$ of the closure Ω^a of Ω in $\hat{M}$ and the ideal boundary $\hat{M} \setminus M$ is not empty. Denote by ∂M the ideal boundary $\hat{M} \setminus M$.

(ii) For some positive number ϵ, there esists $(L_h + \epsilon I)$-Green's function $g_\Omega^\epsilon(x, y)$ on Ω, where L_h is the Laplace-Beltrami operator of h.

(iii) For $x \in M$, $d(x, y) \to +\infty$ as $y \to \partial M$.

(iv) (M, h) has a Green's function $G(x, y)$ of L_h on M such that for every $x \in M$, $G(x, y) = o(1)$ as $y \to \partial_\Omega M$.

There are many manifolds which posses the condition (α). Among these manifolds are

1) Cartan-Hadamard manifolds with bounded negative sectional curvatures, where one can take Ω as the whole manifolds (Ancona [1]) and the Eberlein and O'Neill compactification is the desired one, and

2) Complete Kähler manifolds (D, g). Indeed we have the following

Proposition 1 *There exists a compact set K of D such that (D, g) satisfies the condition (α), where we take as Ω the set $D \setminus K$ and as $\hat{D}$ the topological compactification of D in the Euclidean space $\mathbf{C}^n$, $N = 2n$.*

PROOF. The property (A) is guranteed by the fact that sectional curvatures of (D, g) are asymptotically bounded by two negative constants (cf. [8]). The condition (B)(iv) is an immediate consequence of the Malliavin's estimate of the Green's functions ([11]):

(1) $$G(x, y) \approx dist(y, \partial D)^n, \ for \ x \in D$$

(B)(iii) is obvious. Therefore we need to show (B)(i) and (ii). For this purpose we provide

Lemma 1 *From now on we denote by L_g the Laplace-Beltrami operator of g. Then*

(i) $$L_g(-\lambda)^n = O(\delta) \cdot (-\lambda)^n, \ where \ \delta(z) = dist(z, \partial D).$$

(ii) *For every $0 < \beta < n$, there exists $\epsilon > 0$ such that*

$$L_g(-\lambda)^\beta \leq -\epsilon(-\lambda)^\beta, \ near \ \partial D.$$

(iii) *For every $n < \beta$, there exists $\epsilon > 0$ such that*

$$L_g(-\lambda)^\beta \geq \epsilon(-\lambda)^\beta, \ near \ \partial D.$$

PROOF OF LEMMA 1. Since g is a Kähler metric, its Laplace-Beltrami operator L_g is written as

$$L_g f(z) = 4 \sum g^{\bar{i}j}(z) \frac{\partial f(z)}{\partial \bar{z}_i \partial z_j} \quad z \in D,$$

where $(g^{\bar{i}j}(z))$ is the inverse matrix of the metric $g(z) = (g_{i\bar{j}}(z))$. Let us recall the boundary behavior of $g^{\bar{i}j}$: It is easy to check that there exists an open set U such that $\partial D \subset U$ and for every $z \in U$ there is the unique boundary point $b(z) \in \partial D$ with $|z - b(z)| = \delta(z)$, where $\delta(z)$ is the Euclidean distance between z and ∂D. Let $D_0 = D \cap U$. For $z \in D_0$, we denote by N_z the complex linear space spanned by the inward unit normal $\nu_{b(z)}$, and by C_z the complex tangent space $T_{b(z)}^{\mathbf{C}}(\partial D)$. Then $\mathbf{C}^n = N_z \oplus C_z$. Now we introduce for $z \in D_0$ the following coordinate system:

$$b(z) = 0, \quad N_z = \{(w_1, 0, \cdots, 0) : w_1 \in \mathbf{C}\}, \quad C_z = \{(0, w_2, \cdots, w_n) : (w_2, \cdots, w_n) \in \mathbf{C}^{n-1}\}.$$

Here we take $\nu_{b(z)} = (1, 0, \cdots, 0)$. By [13], near the boundary $(g^{\bar{i}j})$ behaves as follows:

$$(g^{\bar{i}j}) = \lambda^{-2} \begin{pmatrix} \frac{4}{a^2} + O(\lambda) & O(1) \\ \hline O(1) & (-\lambda)^{-1} Q_2^{-1} + O(1) \end{pmatrix},$$

where

$$Q_2 = \left(\frac{\partial \lambda}{\partial z_i \partial \bar{z}_j} \right) \quad and \quad a = |\nabla \lambda(b(z))|.$$

By using this fact we will prove Lemma 1. Let $f = (-\lambda)^n$. For a smooth function u, let

$$u_i = \frac{\partial u}{\partial z_i}, \quad u_{\bar{j}} = \frac{\partial u}{\partial \bar{z}_j}, \quad \text{and} \quad u_{i\bar{j}} = ((u)_i)_{\bar{j}}.$$

Then

$$
\begin{aligned}
L_g f(z) &= \sum_{i,j=1}^{n} (-\lambda)^n \left\{ n(n-1)(-\lambda)^{-2} g^{i\bar{j}}(-\lambda)_i(-\lambda)_{\bar{j}} + n(-\lambda)g^{i\bar{j}}(-\lambda)_{i\bar{j}} \right\} \\
&= : (I) + (II)
\end{aligned}
$$

Since $-\lambda = aImz_1 + O((-\lambda)^2)$, we have $(-\lambda)_1 = a/2 + O(-\lambda)$ and $(-\lambda)_j = O(-\lambda)$, $j \geq 2$. Therefore,

$$
\begin{aligned}
(I) &= n(n-1)(-\lambda)^n \left\{ \left(\frac{4}{a^2} + O(-\lambda) \right) \left(\frac{a^2}{4} + O(-\lambda) \right) + O(-\lambda) \right\} \\
&= \{n(n-1) + O(-\lambda)\} \cdot (-\lambda)^n, \quad \text{and}
\end{aligned}
$$

$$
\begin{aligned}
(II) &= (-\lambda)^n \cdot n(-\lambda)^{-1}(-\lambda)^2 \{O(1) + O(-\lambda) - (n-1)(-\lambda)^{-1}\} \\
&= (-\lambda)^n \cdot n\{O(-\lambda) + O((-\lambda)^2) - (n-1)\} \\
&= \{-n(n-1) + O(-\lambda)\} \cdot (-\lambda)^n
\end{aligned}
$$

Consequently, we have *(i)*.

To prove *(ii)* and *(iii)*, compute $L_g(f^\alpha)$ by using *(i)*. Then we obtain that

$$L_g(f^\alpha) = f^\alpha \{n^2 \alpha(\alpha - 1) + O(-\lambda)\},$$

when $\alpha \neq 1$. This equality implies the desired inequalities. END OF PROOF OF LEMMA 1.

From Lemma 1 (ii) we can take an open set V in $\mathbf{C}^n$ such that $\partial D \subset V$ and $(-\lambda)^{n/2}$ is a positive $(L_g + \epsilon I)$-superharmonic function on $V \cap D$ for a suitable number $\epsilon > 0$ which is not $(L_g + \epsilon I)$-harmonic on $V \cap D$. Hence there exists a positive $(L_g + \epsilon I)$-potential on $V \cap D$, and therefore by Hervé [6] an $(L_g + \epsilon I)$-Green's function on $V \cap D$ exists. Thus we can take as K the compact set $D \setminus (V \cap D)$. ∎

Since g is the main part of the Bergman metric, we can prove likewise that
3) the manifold (D, g_b)
posesses the condition (α).

§2. Boundary Harnack principle for manifolds with condition (α).
In this section, let (M, h) be an N-dimensional Rimannian manifold satisfying the condition (α),

and let r_0, c_0, ϵ and Ω be as in the Definition 1. The main theorem in this section is the so-called boundary Harnack principle which is an analogue of the classical one. We will formulate the principle adapted to the geometry of (M, h). In order to describe its statement we shall extend the notion of Φ-chain defined by Ancona ([1]):

Definition 2 *For a set $E \subset M$, ∂E and $\overline{E}$ represent the boundary and the colsure of E in the topology of M respectively.*

Let V be an open set of M, and let $\Phi : [0, +\infty) \to (0, +\infty)$ be an increasing function with $\lim_{t \to +\infty} \Phi(t) = +\infty$. Put $c_1 = \Phi(0)$. Then the pair $(\Gamma_i, x_i)_{i=1,\cdots,m}$ of open sets $\Gamma_i \subset V$ and points $x_i \in \partial \Gamma_i \cap V$ $(i = 1, \cdots, m)$ is said to be V-Φ-chain if

(i)
$$V \supset \Gamma_1 \supset \cdots \supset \Gamma_m$$

(ii)
$$\frac{1}{c_1} \le d(x_i, x_{i+1}) \le c_1, \text{ and}$$

(iii) *For $z \in \partial \Gamma_i \cap V$,*
$$i = 1, \cdots, m - 1 \implies d(z, \Gamma_{i+1}) \ge \Phi(d(z, x_i))$$
$$i = 1 \text{ or } m \implies d(z, \partial V) \ge \Phi(d(z, x_i)).$$

Our boundary Harnack principle is stated in terms of V-Φ-chain. In order to mention it, we use the following notation:

For $E \subset M$, let $\partial_a E = E^a \setminus E^i$, where E^a and E^i are the closure and the interior of E in the topology of $\hat{M}$ respectively.

Theorem 1 *Let (M, h) and Ω be as in Definition 1. Suppose that V, Ω_{-1}, Ω_0 and Ω_1 are open sets satisfying the following (i) $\sim$ (iii)*

(i) $\Omega \supset V \supset \overline{\Omega_{-1}} \supset \Omega_{-1} \supset \overline{\Omega_0} \supset \Omega_0 \supset \overline{\Omega_1} \supset \Omega_1$.

(ii) There exists $A \in \partial \Omega_0$ such that for some $c_2 > 0$, $\overline{B(A, c_2)} \subset \Omega_{-1} \setminus \Omega_1$.

(iii) Moreover, each point in $\partial \Omega_{-1}$ can be joined to each point in $\partial \Omega_1$ by an V-Φ-chain, through the point A.

Let $S = (\partial M \cap \partial_a V) \cup (\Omega_{-1} \cap \partial_a V)$. If two functions u and v are posititve L_h-harmonic on V and u vanishes continuously at S, then

$$\frac{u(x)}{u(A)} \le c \frac{v(x)}{v(A)} \text{ for all } x \in \Omega_1,$$

where c is a positive constant depending only of M, c_2 and quantity related to the definition of V-Φ-chain.

This theorem is a refinement of Ancona [1, Theorem 5′]. In fact, we have a crucial case to which we can not apply [1, Theorem 5′], but which satisfies all assumptions in Theorem 1. That case will essentially appear when we study regularity of L_g-harmonic measure. Theorem 1 is proved along the lines suggested by A. Ancona in [1].

Here we give two examples satisfying all hypoteses of Theorem 1. The former is involved in [1, Thoerem 5′], but the later does not:

Examples. Let M be the open unit ball and h be a Riemannian metric on M which is asymptotically quasi-isometric to the hyperbolic metric of M. Suppose $\hat{M}$ is the closed unit ball. Let $I = \{x \in \partial M : |x - \zeta| < \delta\}$ for $\zeta \in \partial D$ and a small positive number δ. Denote by kI the surface ball having the same center as I but whose radius is k times as large. Consider the Carleson box $S(J)$ over a surface ball J, that is, $S(J) = \{rx : 1 - r(J) < r < 1, x \in J\}$, where $r(J)$ is the radius of J. Our first example is

(i) Let $V = S(4I)$, $\Omega_{-1} = S(3I)$, $\Omega_0 = S(2I)$ and $\Omega_1 = S(I)$. Then $S = 4I$.
The second example is

(ii) Let $V = S(4I) \setminus \overline{S((1/2)I)}$, $\Omega_{-1} = V \setminus \overline{S(I)}$, $\Omega_0 = V \setminus \overline{S(2I)}$ and $\Omega_1 = V \setminus \overline{S(3I)}$. Then $S = \partial_a V \setminus \partial S((1/2)I)$.

Because of Theorem 1, we have by the same way as in [1] a sufficient condition in order that a point of the ideal boundary is a minimal Martin boundary point (cf. Theorem 2 which will be stated later). Before stating the theorem, we recall the definition of the Martin boundary and some fundamental facts. Fix a point p in M and define the function

$$K(x, y) = \begin{cases} \dfrac{G(x, y)}{G(p, y)} & y \neq p \\ 0 & y = p, x \neq p \\ 1 & x = y = p. \end{cases}$$

Consider a sequence $\sigma = \{y_i\}$ in M having no limit points in M. The sequence is called fundamental if $\{K(x, y_i)\}$ converges to a harmonic function $K_\sigma(x)$ on M. By the interior Harnack inequality, any sequences in M having no limit points in M has a fundamental subsequence. Denote by Σ the set of all fundamental sequences. Given two elements σ and τ of Σ, define $\sigma \sim \tau$ iff $K_\sigma = K_\tau$ on M. Then $\sim$ is an equivalence relation in Σ. The Martin boundary Δ is defined as the set of equivalence classes $\Sigma/\sim$. For $[\{y_i\}] \in \Delta$, let $K(x, [\{y_i\}]) = \lim_{i \to \infty} K(x, y_i)$. Put $\mathcal{M} = M \cup \Delta$. Let

$$\varrho(y, z) = \int_{B(p,1)} \frac{|K(x, y) - K(x, z)|}{1 + |K(x, y) - K(x, z)|} dv_h(x), \quad y, z \in \mathcal{M}$$

where dv_h is the Riemannian measure induced by the metric h. Then the space $(\mathcal{M}, \varrho)$ is compact and its topology inside of $\mathcal{M}$ agrees with the initial topology of M. The set Δ is called the Martin boundary, and

$$\Delta_1 = \{\sigma \in \Delta : K(x, \sigma) \text{ is a minimal } L_h \text{harmonic function on } M\}$$

is called the minimal Martin boundary of M.

Now we proceed to give our criterion. By the same way as in [1], we can define a modified (G.A.) condition:

Definition 3 *A point $\zeta \in \partial M$ is said to satisfy (G.A.) if there exists a basis of neighborhood $\{V_k(\zeta)\}$ of ζ with respect to the topology of $\hat{M}$ such that*

(i) $U_k(\zeta) := V_k(\zeta) \cap M \subset \Omega$,

(ii) $U_k(\zeta) \neq \phi$, $U_k(\zeta) \supset \overline{U_{k+1}(\zeta)}$,

(iii) there exists $A_k \in U_{2k}(\zeta) \setminus \overline{U_{2k+1}(\zeta)}$ such that each point of $\partial U_{2k}(\zeta)$ can be joined to each point in $\partial U_{2k+1}(\zeta)$ by an Ω-Φ-chain, through A_k, and

(iv) $\bigcap_{k=1}^{\infty} U_k(\zeta) = \phi$.

Let $\partial_1 M = \{\zeta \in \partial M : \zeta \text{ satisfies (G.A.)} \}$.

Let us call such sequence $\{V_k(\zeta)\}$ a Harnack chain to ζ.

The following is proved by a similar way as in [1], because we have obtained Theorem 1:

Lemma 2 *For $\zeta \in \partial_1 M$, there exists the unique point $B(\zeta)$ in Δ_1 such that any sequence $\{x_j\}$ in M converges to $B(\zeta)$ in the topology of $\mathcal{M}$ if and only if for each $k \geq 1$, $x_j \in U_k(\zeta)$ for j large enough.*

By this lemma we can easily gain the following

Theorem 2 . *(i) The mapping $B : \partial_1 M \to \Delta_1$ is a continuous injection.*

(ii) If $\partial_1 M = \partial M$, then B is a homeomorphism from ∂M onto Δ_1.

PROOF. Since (i) is an immediate consequence of Lemma 2, we prove here (ii). It suffices to prove that B is surjective. Take an arbitrary point σ in Δ_1. Then there exists by its definition a fundamental sequence $\{y_i\}$ in M with $\varrho(y_i, \sigma) \to 0$ as $i \to 0$. Hence it is a Cauchy sequence with respect to the distance ϱ, and therefore one with respect to the initial topology of M. Consequently, there exists a limit point $\zeta \in \partial_1 M$ of $\{y_i\}$. Again by Lemma 2 we get $B(\zeta) = \sigma$.∎

Lemma 2 can be apply also to the Dirichlet problem:

Theorem 3 *Suppose $\partial_1 M$ is closed and B is homeomorphism from $\partial_1 M$ onto Δ_1. Then for any countinuous function f on $\partial_1 M$, there exists the unique continuous function $H(f)$ on $M \cup \partial_1 M$ such that $H(f)$ is L_h-harmonic on M and $H(f)(\zeta) = f(\zeta)$ for all $\zeta \in \partial_1 M$.*

From this theorem it follows that when all assumptions of Theorem 3 are satisfied, then for $z \in M$ there exists the unique probability Radon measure ω^z on $\partial_1 M$ such that for any continuous function f on $\partial_1 M$,

$$H(f)(z) = \int_{\partial_1 M} f(\zeta) d\omega^z(\zeta).$$

Such measure ω^z is called the L_h-harmonic measure relative to z.

§3. Construction of Harnack chains in (D, g).

In this section we will construct a Harnack chain to every point of the boundary of the strictly pseudoconvex domain D. First, for investigating the geometric structure of a neighborhood of the boundary, we will deform the Kähler metric g. Since the Harnack chains are invariant under quasi-isometric deformation, we can deform quasi-isometrically the metric g in order to construct them.

For a point $\zeta \in \partial D$, take a coordinate neighborhood (W_ζ, φ_ζ) (abbraviate (W_ζ, φ)) satisfying

(i) W_ζ is an open set in $\mathbf{C}^n$ which contains ζ,

(ii) φ is diffeomorphism from W_ζ onto the open unit ball $B_e(0, R) \subset \mathbf{C}^n$ with center 0 and radius R with respect to the Euclidean metric of $\mathbf{C}^n$, $\varphi_\zeta(\zeta) = 0$,

(iii)
$$\varphi : D \cap W_\zeta \rightarrow \{z \in B_e(0, R) : Rez_1 > 0\},$$
$$\varphi : \partial D \cap W_\zeta \rightarrow \{z \in B_e(0, R) : Rez_1 = 0\},$$
$$\varphi^* \nu_w = \left(\tfrac{\partial}{\partial x_j}\right)_{\varphi(w)}, \quad w \in \partial D \cap W_\zeta, \text{ where } \nu_w \text{ is the inward unit normal}$$
vector on ∂D at w, $z = (z_1, \cdots, z_n)$ and $z_j = x_{2j-1} + \sqrt{-1} x_{2j}$,

(iv) for an orthonormal frame fields $\kappa_j, j = 3, \cdots, 2n$ of $T^C(\partial D)$ on $\partial D \cap W_z$,

$$(\varphi^* \kappa_j)_{\varphi(\zeta)} = \left(\frac{\partial}{\partial x_j}\right)_0, \quad and$$

(v)
$$|\nabla \varphi| \leq A,$$
where A is a constant depending only on D.

Let $X_1 = \varphi^* \nu$, $X_2 = \varphi^* \sqrt{-1} \nu$, $X_j = \varphi^* \kappa_j$, $j = 3, \cdots, 2n$. Then $X_1, \cdots, X_{2n}$ are frame fields of $T(\{0\} \times \mathbf{R}^{2n-1})$ on $B_e(0, R)$ (but not necessarily orthogonal).

Consider the following metric on $H = (0, +\infty) \times \mathbf{R}^{2n-1}$:

For $\xi = \sum_{j=1}^{2n} \xi^j X_j$, $\eta = \sum_{j=1}^{2n} \eta^j X_j \in T\mathbf{R}^{2n}$, let $\bar{g}_x(\xi, \eta) = \bar{g}_{ij}(x)\xi^i \eta^j$ for $x \in B_e(0, r)$,

where

$$\bar{g}_{11} = \bar{g}_{22} = \frac{1}{(x_1)^2}, \ \bar{g}_{jj} = \frac{1}{x^1}, \ j = 3, \cdots, 2n, \ \bar{g}_{ij} = 0, \ i \neq j.$$

We get the following

Theorem 4 *There exists an open neighborhood D_0 of ∂D such that for every $\zeta \in \partial D$, the pull back $\tilde{g}$ of the metric $\bar{g}$ by the coordinate map φ_ζ satisfies that*

$$\frac{1}{C}\tilde{g}_w(\xi,\xi) \leq g_w(\xi,\xi) \leq C\tilde{g}_w(\xi,\xi)$$

and

$$\frac{1}{C}\tilde{g}_w(\xi,\xi) \leq (g_b)_w(\xi,\xi) \leq C\tilde{g}_w(\xi,\xi)$$

for $w \in W_\zeta \cap D \cap D_0$ and $\xi \in T_w(D)$, where C is a positive constant depending only on D, g and the constant A appeared in the definition of φ_ζ. (Note that A is independent of ζ).

By using this theorem we construct Harnack chain for (D,g) as follows: For $x \in H \cap B_e(0,R)$, let

$$\overline{T}[x] = \{y \in B_e(0,R) : 0 \leq y_1 < x_1, \ \bar{g}_x((0,y'),(0,y')) < e^{-\delta}\},$$

where $y' = (0,y_2,\cdots,y_{2n})$ and δ is a suitable large positive number. For $\zeta \in \partial D$, let $\zeta(r) = \zeta + r\nu_\zeta, r > 0$. Let

$$T[\zeta,r,\varphi_\zeta](= T[\zeta,r]) = \varphi_\zeta^{-1}(\overline{T}[\varphi_\zeta(\zeta(r))]).$$

Then we obtain

Theorem 5 *There exists a small positive number ε_0 such that for $\zeta \in \partial D, T[\zeta,\varepsilon_0^k,\varphi_\zeta](k = 1,2,\cdots)$ are Harnack chain to ζ with respect to the metric g and the Bergman metric g_b.*

From this theorem, Theorem 2 and 3 we have

Corollary 1 *(i) ∂D is homeomorphic to the minimal Martin boundary of (D,g).*

(ii) The Dirichlet problem for L_g and ∂D is solvable.

(iii) Let h be a positive L_g-harmonic function on D. Then there exists the unique positive Borel measure μ_h on ∂D such that for any $z \in D$,

$$h(z) = \int_{\partial D} K(z,\zeta)d\mu_h(\zeta).$$

The measure μ_h is called the Martin representing measure of h.

(iv) For $z \in D$, the L_g-harmonic measure ω^z relative to z and (D,g) exists.

The statements (i) - (iv) also hold true for the Bergman metric g_b.

Remark. This corollary gives the affirmative answer to a problem raised by J. C. Taylor (p.244 of [15]).

§4. Comparison of the bottom of $T[\zeta, r]$ and a non-isotropic ball. For further study of the strictly pseudoconvex Kähler domain (D, g), we will mention about form of the bottom of $T[\zeta, r]$ for $zeta \in \partial D$ and $r > 0$. Denote by $\Delta(\zeta, r)$ the bottom of $T[\zeta, r]$, that is, $T[\zeta, r] \cap \partial D$. Let us recall the definition of non-isotropic balls: For $\zeta \in \partial D$ and $r > 0$, the non-isotropic ball $\beta(\zeta, r)$ with center ζ and radius r is defined by

$$\beta(\zeta, r) = \{\eta \in \partial D : |(\zeta - \eta) \cdot \nu_\zeta| < r, \ |\zeta - \eta|^2 < r\}.$$

Our result is the following

Theorem 6 *For $\zeta \in \partial D$ and $r > 0$,*

$$\Delta(\zeta, c^{-1}r) \subset \beta(\zeta, r) \subset \Delta(\zeta, cr),$$

where c is a constant depending only on D, g and A.

As an application of this theorem, we will give new Harnack chains constructed from non-isotropic balls. It is easy to check that there exists an open set U in $\mathbf{C}^n$ such that $\partial D \subset U$ and for every $z \in U$ there is the unique boundary point $b(z) \in \partial D$ with $|z - b(z)| = \delta(z)$, where $\delta(z)$ is the Euclidean distance between z and ∂D, that is, $dist(z, \partial D)$.

Corollary 2 *For $\zeta \in \partial D$ and $r > 0$, let*

$$C(\zeta, r) = \{z \in D \cap U : 0 \le \delta(z) < r, \ b(z) \in \beta(\zeta, r)\}.$$

Then $\{C(\zeta, 2^{-k})\}$ is a Harnack chain to ζ for sufficient large k.

In the next section we will use this Harnack chain.

All results in this section are correct for the Bergman metric g_b.

§5. Estimates of L_g-harmonic measure. In this section we will study L_g-harmonic mesures. For the reference point p defined in section 2, let $\omega = \omega^p$. By the bounded geometry property of (D, g) and the interior Harnack inequality (Theorem 8.20 in [5]), two L_g-harmonic meaures $d\omega$ and $d\omega^z$, $z \in D$, are mutually absolutely continuous, and

$$K(z, \zeta) = \frac{d\omega^z}{d\omega}(\zeta), \ \zeta \in \partial D,$$

where the fractional is the Radon-Nikodym derivative.

One of our main theorems is the following

Theorem 7 *Let $d\sigma$ be $(2n - 1)$-dimensional area Hausdorff measure on ∂D. Then $d\sigma$ and $d\omega$ are mutually absolutely continuous. Moreover, there exists a function k on ∂D such that*

$$k, \ \frac{1}{k} \in L^\infty(\partial D, d\sigma), \ and \ d\omega = kd\sigma.$$

From this theorem we can regard $P(z, \zeta) = K(z, \zeta)k(\zeta)$ as an analogue of the Poisson-Szegö kernel for the strictly pseudoconvex domain D.

The main tool for proving this theorem is the boundary Harnack's principle associated with the chain $\{C(\zeta, r)\}$, that is,

Lemma 3 *Let $\zeta \in \partial D$ and $r > 0$.*

(i) Let u and v be positive L_g-harmonic function on $C(\zeta, 4r) \cap D$. Suppose u vanishes continuously on $\overline{C}(\zeta, 4r) \cap \partial D$. Then

$$\frac{u(z)}{u(\zeta(2r))} \leq c \frac{v(z)}{v(\zeta(2r))},$$

for all $z \in C(\zeta, r) \cap D$, where c is a constant depending only on D and g.

(ii) Let u and v be positive L_g-harmonic functions on $[C(\zeta, 8r) \setminus C(\zeta, r)] \cap D$. Suppsoe u vanishes continuously on $\partial_a(C(\zeta, 8r) \setminus C(\zeta, r))$. Then

$$\frac{u(z)}{u(\zeta(2r))} \leq c \frac{v(z)}{v(\zeta(2r))},$$

for all $z \in [C(\zeta, 8r) \setminus C(\zeta, 4r)] \cap D$.

We need also the next lemma:

Lemma 4 *Tere exists a positive constant c such that for every $\zeta \in \partial D$ and $r > 0$,*

$$\omega^{\zeta(r)}(\beta(\zeta, r)) \geq c > 0.$$

The main part of the proof of Theorem 7 is the following

Lemma 5 *There exists a positive constant C depending only on D and g such that*

$$\frac{1}{C}\sigma(\beta(\zeta, r)) \leq \omega(\beta(\zeta, r)) \leq C\sigma(\beta(\zeta, r)),$$

for all $\zeta \in \partial D$ and $r > 0$.

OUTLINE OF PROOF OF THEOREM 7. Consider the Lebesgue decomposition, $d\omega = d\omega_a + d\omega_s$, of $d\omega$ with respect to $d\sigma$, where $d\omega_a$ and $d\omega_s$ are the absolutely continuous part and the singular part of $d\omega$ respectively. From Lemma 5 it follows that for all $\zeta \in \partial D$ and $r > 0$,

(∗).
$$\frac{\omega_s(\beta(\zeta, r))}{\sigma(\beta(\zeta, r))} \leq C$$

Now, this inequality and the following lemma we get that $d\omega_s = 0$:

Lemma 6 *Let X be a compact topological space. Suppose that a function $\rho : X \times X \to [0, +\infty]$ satisfies*

(i) for every $x \in X$ and $r > 0$, the set $B_\rho(x,r) = \{y \in X : \rho(x,y) < r\}$ is open,

(ii) $\rho(x,y) = 0$ if and only if $x = y$, and

(iii) there exists a positive constant A_1 such that for $x, y, z \in X$,

$$\rho(x, y) \leq A_1[\rho(x, z) + \rho(y, z)].$$

Let μ be a positive Borel measure on X. If μ is the doubling measure with respect to ρ, that is, there exists a constant $A_2 > 0$ so that

$$\mu(B_\rho(x, 2r)) \leq A_2\mu(B_\rho(x, r)),$$

for any $x \in X$ and $r > 0$. If a nontrivial positive Borel measure ν is singular with respect to μ, then

$$\lim_{r \to 0} \frac{\nu(B_\rho(x, r))}{\mu(B_\rho(x, r))} = +\infty \quad for \quad \nu\text{-a.e. } x.$$

Apply this lemma with $X = \partial D$, $\nu = \omega_s$, $\mu = \sigma$ and

$$\rho(x, y) = \inf\{\delta > 0 : y \in \beta(x, \delta)\},$$

for $x, y \in \partial D$. Then $B_\rho = \beta$, and therefore we conculude that $d\omega_s$ must be trivial because of (*).

Since we have by Lemma 4 that ω is the doubling measure with respect to ρ, we can apply Lemma 6 with $X = \partial D$, $\mu = \omega$ and $\nu = \sigma_s$, where σ_s is the singular part of σ with respect to ω, and from Lemma 5 it follows that $\sigma_s = 0$. Therefore σ and ω are mutually absolutely continuous, and by Lemma 5 we get that there exists the desired function k. ∎

§6. Boundary behavior of L_g-harmonic functions. In this senction we will apply our results to the study of the boundary behavior of L_g-harmonic functions. In fact, we will extend the local version of the relative Fatou theorem for the unit ball obtained by Koranyi and Taylor [14] and [9] to one for the strictly pseudoconvex domain D.

The well known classical Fatou theorem states that a positive harmonic function associated with the Euclidean metric of $\mathbf{C}^n$ converges nontangentially at $d\sigma$-almost every boundary point. However, A. Koranyi discovered that a positive L_g-harmonic function f for the open unit ball converges admissibly at $d\sigma$-a.e. boundary point, that is, for any $\alpha > 1$, the limit

$$\lim_{z \to \zeta, z \in A_\alpha(\zeta)} f(z)$$

exists $d\sigma$-a.e. $\zeta \in \partial D$, where

$$A_\alpha(\zeta) = \{w : |w| < 1, |1 - w \cdot \bar{\zeta}| < \alpha(1 - |w|)\}.$$

A. Koranyi obtained a local version of this result, the so-called local Fatou theorem. Local relative Fatou theorem is a refinement of this theorem.

A crucial difference between "non-tangentially" and "admissibly" is that the domain $A_\alpha(\zeta)$ contains a $(2n-2)$-dimensional submanifold which tangents to the boundary at ζ. Hence positive L_g-harmonic functions have certain tangential boundary limits at almost every point of the boundary. This phenomena does not happen in the case of positive harmionic functions with respect to the Euclidean metric.

After this theorem E. M. Stein extended the local Fatou theorem for the unit ball to the domain D, but he assumed the considered functions are holomorphic. We here prove a complete extension of the local relative Fatou theorem for the domain D. Before stating our theorem, let us recall that analogue of the admissible domain $A_\alpha(\zeta)$ for the strictly pseudoconvex domain which was obtained by E. Stein [13] :

For $\alpha > 1$ and $\zeta \in \partial D$, let

$$A_\alpha(\zeta) = \{z \in D : |(z - \zeta) \cdot \nu_\zeta| < \alpha\delta(z), \ |z - \zeta|^{\frac{1}{2}} < \alpha\delta(z)\},$$

where $\delta(z) = dist(z, \partial D)$. And a function on D is said to converge admissibly to the point ζ if for any $\alpha > 1$, the limit

$$\lim_{z \to \zeta, z \in A_\alpha(\zeta)} f(z)$$

exists.

Now we can state our main theorem in this section :

Theorem 8 *Let u and h be L_g-harmonic functions on D, and let E be a subset of ∂D. Suppose that h is positive on D and that there exists a number $\alpha > 1$ satifying*

$$\inf_{z \in A_\alpha(\zeta)} \frac{u(z)}{h(z)} > -\infty,$$

for every $\zeta \in E$. Then there exist subsets F_1 and F_2 of ∂D so that $\sigma(F_1) = \mu_h(F_2) = 0$, where μ_h is the Martin representing measure of h, and u/h converges admissibly at every point of the set $E \setminus (F_1 \cup F_2)$.

We note that Theorem 7 and Theorem 8 also hold true for g_b.

In view of our results it seems justified to call attention to study admissibly area integral functions and admissibly maximal functions of the Bergman harmonic functions: For a function f on D and $\zeta \in \partial D$, let

$$\mathbf{A}_\alpha(f)(\zeta)^2 := \int_{A_\alpha(\zeta)} \| gradf(z) \|^2 \, dv(z), \quad \text{and}$$

$$\mathbf{N}_\alpha(f)(\zeta) := sup_{z \in A_\alpha(\zeta)}|f(z)|,$$

where $\| \, gradf(z) \, \|$ is the norm of the gradient of f with respect to the metric g_b or g, and dv is the Riemannian measure for g_b or g. The following problem is naturally raised:

Problem 1. Study harmonic analysis for $\mathbf{A}_\alpha(f)$ and $\mathbf{N}_\alpha(f)$.

Now, let k be the function appeared in Theorem 7. We close this paper with posing the following problems:

Problem 2. Is k smooth on ∂D ?

Problem 3. Is the Martin kernel $K(z, \zeta)$ for the Bergman metric or the metric g smooth on ∂D ?

Detailed proofs of the results in this paper will appear in later papers.

References

[1] A. Ancona, Negatively curved manifolds, elliptic operators, and the Martin boundary, Ann of Math., 125 (1987), 495-536.

[2] M. Anderson and R. Schoen, Positive harmonic functions on complete manifolds of negative curvature, Ann of Math., 121 (1985), 429-461.

[3] H. Arai, Estimates of harmonic measures associated with degenerate Laplacian on strictly pseudoconvex domains, Proc. Japan Acad., 66, Ser.A, (1) (1990), 13-15.

[4] C. Fefferman, The Bergman kernel and biholomorphic mappings of pseudoconvex domains, Invent. Math., 26 (1974), 1-65.

[5] D. Gilberg and N. S. Trudinger, Elliptic Partial Differential Equations of Second Order, 2nd edition, Springer-Verlag, Berlin, Heidelberg, New York Tokyo, 1984.

[6] R.-M. Hervé, Recherches axiomatiques sur la théorie des fonctions surharmoniques et du potentiel, Ann. Inst. Fourier, Grenoble 12 (1962), 415-571.

[7] S. Ito, Fundamental solutions of parabolic differential equations and boundary value problems, Japan. J. Math., 27 (1957), 55-102.

[8] P. F. Klembeck, Kähler metrics of negative curvature, the Bergman metric near the boundary, and the Kobayashi metric on smooth bounded pseudoconvex sets, Indiana Univ. Mathe. J., 27 (1978), 275-282.

[9] A. Korányi and J. C. Taylor, Fine and admissible convergence for symmetric spaces of rank one, Trans. Amer. Math. Soc., 263 (1981), 169-181.

[10] S. G. Krantz, Function Theory of Several Complex Variables, John Wiley & Sons, New York, 1982.

[11] P. Malliavin, Fonctions de Green d'un ouvert strictement pseudo-convexe et classe de Nevanlinna, C. R. Acad. Sci. Paris, 278 (1974), 141-144.

[12] T. Sasaki, On the Green functioin of a complete Remannian or Kähler manifold with asymptotically negative constant curvature and applications, Advanced Studies in Pure Math., 3 (1984), 387-421.

[13] E. M. Stein, Boundary Behavior of Holomorphic Functions of Several Complex Variables, Princeton Univ. Press 1972.

[14] J. C. Taylor, Fine and admissible convergence for the unit ball in $\mathbf{C}^n$ (Proc. 1979 Copenhagen Potential Theory Colloquium, 1979), Lect. Notes in Math., vol 787, Springer-Verlag, Berlin and New York (1980), 299-315.

[15] Potential Theory, Surveys and Problems, Proceedings, Prague 1987 (J. Král, J. Lukeš, I. Netuka and J. Veselý (Eds.)), Lect Notes in Math., vol.1344, Springer-Verlag, Berlin and New York, 1987.

Uniqueness and non uniqueness for harmonic functions with zero nontangential limits

J. Marshall Ash*, DePaul University, Chicago, IL 60614

Russell Brown†, University of Kentucky, Lexington, KY 40506

1. Introduction.

Definitions. By D we mean the open unit disc which is centered at the origin in the complex plane and by T we mean its boundary, *i.e.*, its circumference.

The question we will address here is to what extent does the limiting behavior of a harmonic function on T determine its values on D. A simple result in this direction follows from the maximum principle: If a harmonic function has limit 0 at each point of T, then the function is 0 on D (see Thm. 1 in Section 2). Emboldened by this, one might conjecture that if a harmonic function merely has radial limit 0 at each point of T then the function is 0 on D. Unfortunately, this is not so. In fact, consider the function $u_1(r, \theta) := Im\left(\frac{z}{(1-z)^2}\right) = \sum_{n=1}^{\infty} n \sin(n\theta)r^n$, which is harmonic on D. If $e^{i\theta} \neq 1$,

$$\lim_{z \to e^{i\theta}} u_1(z) = Im\left(\frac{e^{i\theta}}{(1-e^{i\theta})^2}\right) = Im(\frac{1}{4} \csc^2 \frac{\theta}{2}) = 0,$$

and $\lim_{r \to 1^-} u_1(r, 0) = \lim_{r \to 1^-} \sum 0 = 0$.

Although u_1 is unpleasant, it is, in a strong sense, the worst that can happen. Given a harmonic function u and a positive number r, let $m(r) = m(r, u) := \sup_{|z| \leq r} |u(z)|$. The classical Theorem 3 of Section 2 asserts that if a harmonic function has radial limit 0 at each point of T and if $m(r) = o((1-r)^{-2})$, then the function is 0 on D. That $m(r, u_1)$ is exactly $O((1-r)^{-2})$ is a reflection of the sharpness of this result.

Definition. For any $0 < \alpha < \pi$, let C_α be the circumference $|z| = \sin \frac{\alpha}{2}$. By $\bar{\Omega}_\alpha$ we mean the closed region bounded between the two tangents from $z = 1$ to C_α and by the more distant arc of C_α between the points of contact. Then $\Omega_\alpha := \bar{\Omega}_\alpha \setminus \{1\}$ and the Stolz region $\Omega_\alpha(w)$ is the region Ω_α rotated through an angle $\arg(w)$ around $z = 0$. Note that the angle between the two straight edges of Ω_α is α. When there is no confusion, we shall write $\Omega(w)$ instead of $\Omega_\alpha(w)$.

Definition. We say that the nontangential limit of u at w is s and write

$$\lim n.t._{z \to w} u(z) = s$$

if, for each choice of α, $0 < \alpha < \pi$,

$$\lim_{\substack{z \to w \\ z \in \Omega_\alpha(w)}} u(z) = s.$$

The function u_1 does not have a limit as $z \to 1$ while staying within $\Omega_\alpha(1)$ no matter how small $\alpha > 0$ is chosen, so it seems reasonable to conjecture that if a harmonic

---

* Partially supported by a leave of absence granted by the DePaul University Research Council.

† Supported in part by the National Science Foundation.

function has nontangential limit 0 at every point of T, then the function is 0 on D. Even this is not so. In section 3 we present two examples of nontrivial harmonic functions on D which have nontangential limit 0 at each point of T. The first is somewhat complicated, but has less rapid growth at the boundary. The second was communicated to us by Walter Rudin.

The examples of Section 3 show that some growth condition is necessary. By choosing modes of convergence intermediate between the unconditional limit of Theorem 1 and the radial limit of Theorem 3, and pairing them with corresponding growth rates intermediate between the vacuous one of Theorem 1 and the very restrictive one of Theorem 3, we will interpolate a scale of theorems indexed by a real parameter α between Theorem 1 and Theorem 3. These theorems, Theorem 2α, Section 2, form a more quantitative version of this corollary:

Corollary 1. *If a harmonic function has nontangential limit 0 at each point of T and if there is a real number N so that $m(r) = o((1 - r)^{-N})$ as $r \to 1$, then the function is 0 on D.*

Neither this corollary nor Theorem 2α are as sharp as a result of F. Wolf (see [W], page 65, last sentence and page 66, first sentence) which allows $m(r)$ to be larger, but our method of proof is different from Wolf's.

Although the examples of Section 3 show that some growth condition is necessary for Theorem 2α, they are somewhat disappointing in that they do not give any insight as to whether even the growth rate required by Wolf's version of Theorem 2α is really necessary. A second corollary, Corollary 2, Section 2, applies Theorem 2α to trigonometric series.

2. Results

Theorem 1. *Let u be harmonic on D. If*

$$\lim_{\substack{z \to w \\ |z| < 1}} u(z) = 0$$

for each $w \in T$, then $u(z) = 0$ for all $z \in D$.

Proof. The maximum of a harmonic function which is continuous on the closure of D is attained on T. QED

Theorem 2α. *Let u be harmonic on D. Let $\alpha \in [0, \pi)$. If*

$$\lim_{\substack{z \to w \\ z \in \Omega_\alpha(w)}} u(z) = 0$$

for each $w \in T$, and if

$$m(r) = o \left(\frac{1}{(1 - r)^{\frac{2\pi}{\pi - \alpha}}} \right)$$

then $u(z) = 0$ for all $z \in D$.

Corollary 2γ. Let $\frac{a_0}{2} + \sum(a_n \cos n\theta + b_n \sin n\theta)$ be a trigonometric series satisfying $|a_n| + |b_n| = o(n^\gamma)$, as $n \to \infty$. Assume that for each $w \in T$, (writing $z = re^{i\theta}$) we have

$$\lim_{\substack{z \to w \\ z \in \Omega_\alpha(w)}} \left\{ \frac{a_0}{2} + \sum(a_n \cos n\theta + b_n \sin n\theta)r^n \right\} = 0,$$

where $\alpha = \pi \left(\frac{\gamma - 1}{\gamma + 1} \right)$. Then all a_n and all b_n are 0.

Proof of Corollary 2γ. The sum in curly brackets, call it u, is harmonic in D. It is easy to see that $m(r) = \sum_{n=1}^{\infty} o(n^\gamma)r^n = o\left(\frac{1}{(1-r)^{\gamma+1}} \right)$. (Use formulas III.1.9 and III.1.15 of [Z] for this.) Note that if $\alpha := \pi \left(\frac{\gamma - 1}{\gamma + 1} \right)$, then $\frac{2\pi}{\pi - \alpha} = \gamma + 1$ and apply Theorem 2α with this α to get that $u(r, \theta) = 0$ on D. For any fixed $r < 1$, the series defining $u(\theta) = u(r, \theta)$ converges uniformly to zero. By I.4.10 of [Z], all of the a_n and b_n are 0. QED

Theorem 3. *(F. Wolf [W], V. Shapiro [S], B. E. J. Dahlberg [D], compare S. Verblunsky [Z], IX.8.1, [V])* Let u be harmonic on D. If $\lim_{r \to 1^-} u(rw) = 0$ for each $w \in T$, and if $m(r) = o\left(\frac{1}{(1-r)^2} \right)$, then $u(z) = 0$ for all $z \in D$.

Remark. The various proofs of Theorem 3 have different embellishments, such as allowing small exceptional sets with additional hypotheses. Of course Theorem 2α also admits some of these extensions without much additional effort, but we resist that temptation here.

Remarks. Our proof of Theorem 2α follows Dahlberg's proof of Theorem 3. In fact, if you set $\alpha = 0$ in the proof of Theorem 2α given below, you will have essentially Dahlberg's proof of Theorem 3. [D] Similarly, if you set $\gamma = 1$ in Corollary 2, Verblunsky's uniqueness theorem for Abel summable trigonometric series follows. [Z],IX.8.1, [V] Finally, Theorem 1 may be thought of as Theorem 2α with $\alpha = \pi$.

Definition. To an arc I of T associate the curvilinear triangle $S(I) := \{tw \in D : w \in I$ and $0 \le t < 1\}$. For future reference, we note that $\bar{S}(\bar{I}) = S(\bar{I}) \cup \bar{I}$.

Proof of Theorem 2α. Let u be harmonic on D and fix α in $[0, \pi)$. Assume that

$$\lim_{\substack{z \to w \\ z \in \Omega_\alpha(w)}} u(z) = 0$$

for each $w \in T$. Our goal is to show that $u(z) = 0$ for every $z \in D$. Let $\mathcal{O} := \{w \in T : \limsup_{z \to w} u(z) \le 0\}$. It suffices to show that $\mathcal{O} = T$. For by symmetry it would then be immediate that $\{w \in T : \lim_{z \to w} u(z) \ge 0\} = T$ also, so that $\{w \in T : \lim_{z \to w} u(z) = 0\} = T$. The goal would be reached, since this is exactly the hypothesis of Theorem 1. Letting $u^+(z) := \max\{u(z), 0\}$ as usual, it is clear that $\mathcal{O} = \{w \in T : \lim_{z \to w} u^+(z) = 0\}$. Collecting the known properties of u^+, we will restate what must be done as Lemma 1. Thus, modulo the proof of Lemma 1 we are done. QED

Lemma 1. *If $p(z)$ is*

(1)   subharmonic, continuous, and non-negative on D,

(2) $\lim\limits_{\substack{z \to w \\ z \in \Omega_\alpha(w)}} p(z) = 0$ *for each $w \in T$,*

and if $m(r) := \sup\{p(z) : |z| \leq r\}$ satisfies

(3) $m(r) = o\left(\dfrac{1}{(1-r)^{\frac{2\pi}{\pi-\alpha}}}\right);$

then $\mathcal{O} := \{w \in T : \lim_{z \to w} p(z) = 0\} = T$.

Proof of Lemma 1. We first show that $\mathcal{O}$ is open. Let $w_o = e^{i\theta}$ be a point of $\mathcal{O}$. Then there is a neighborhood of w_o, say of the form $\{(r, \varphi) : 1 - \delta < r < 1$ and $|\varphi - \theta| \leq \delta\}$ for some $\delta > 0$, on which p is bounded. But p, being continuous on the compact set $\{(r, \varphi) : 0 \leq r \leq 1 - \delta$ and $|\varphi - \theta| \leq \delta\}$ is also bounded there. Hence p is bounded on $S(I)$ where I $:= \{e^{i\varphi} \in T : |\varphi - \theta| \leq \delta\}$. To proceed with the proof of Lemma 1 we will need:

Lemma 2. *Let p satisfy (1) and be bounded on $S(I)$ for some closed interval $I \subset T$. Suppose $\lim_{r \to 1^-} p(rw) = 0$ for each $w \in I$, then $\lim_{z \to w} p(z) = 0$ for each w interior to I.*

Proof of Lemma 2. Let $M := \sup\{p(z) : z \in S(I)\}$. Let F be a conformal map of the unit disc onto $S(I)$ and let J be the closed interval of T satisfying $F(J) = I$. Define a function v on D by $v(\zeta) = p(F(\zeta))$. Then v still enjoys property (1) and $0 \leq v \leq M$ on D. Then v has a least harmonic majorant h ([T], pp. 172–173). Since the constant function M is itself a harmonic majorant of v, $h \leq M$. Since h is a bounded harmonic function on D, $h = PI(H)$ for some function H on T. We are using the notation $PI(H)$ to denote the Poisson integral of H:

$$\frac{1}{2\pi} \int_0^{2\pi} H(e^{i\varphi}) \frac{1 - r^2}{1 - 2r \, \cos(\theta - \varphi) + r^2} \, d\varphi.$$

Also, $\lim_{\zeta \to \eta}$ n.t. $h(\zeta) = H(\eta)$ almost everywhere [T], pp. 172-173. Since $\Omega(w)$ contains the radius terminating at w, it follows from (2) that

$$\lim\limits_{\substack{\zeta \to \eta \\ \zeta \in C(\eta)}} v(\zeta) = 0$$

for each η interior to J where for $\eta = F(w)$, $C(\eta) := F^{-1}(\{rw : 0 \leq r \leq 1\})$ is a curve orthogonal to T at η. It follows that H must be 0 almost everywhere on J. In particular, H is essentially 0 on J and hence essentially continuous there, so that $\lim_{\zeta \to \eta} h(\zeta) = 0$ at all points interior to J ([T], p. 130). But then v is squeezed between 0 and h so $\lim_{\zeta \to \eta} v(\zeta) = 0$ everywhere on the interior of J, which is to say that $\lim_{z \to w} p(z) = 0$ everywhere on the interior of I. This proves Lemma 2.   QED

Returning to the proof of Lemma 1, note that since the radius terminating at w is contained in $\Omega(w)$ for all choices of Ω, from Lemma 2 we can conclude that every point of T within δ of w_o is in $\mathcal{O}$. Thus each point of $\mathcal{O}$ is interior to $\mathcal{O}$, so $\mathcal{O}$ is an open subset of T.

Let $p^*(w) := \sup\{p(z) : z \in \Omega(w)\}$. Define $F_j := \{w \in T : p^*(w) \leq j\}$. Then F_j is closed. For if $\{w_k\}$ is a sequence of points in F_j tending to w, and if $z \in \Omega(w)$, then for each k there is a point $z_k \in \Omega(w_k)$ with $|z_k - z| < |w_k - w|$. Since $p(z_k) \leq p^*(w_k) \leq j$ and since p is continuous at z, $p(z) \leq j$. Since z was arbitrary, $p^*(w) \leq j$, $w \in F_j$, so F_j is closed.

Our goal is to show $\mathcal{O} = T$; so, letting $K := T\backslash\mathcal{O}$, we must show the closed set K to be empty. Assume not. From (2) it follows that $\cup F_j = T$, so that $\cup(F_j \cap K) = K$. By the Baire Category Theorem, there is an interval $I \subset T$ and an integer j so that the nonempty set $K \cap I$ is contained in $K \cap F_j$, i.e., $K \cap F_j$ contains a portion of K [Z], I.12.1 If $I = (a, b)$, let $M = \max\{p^*(a), p^*(b), j\}$. To prove Lemma 1, we will show

(4) $$p(z) \leq M \quad \text{for every } z \in S(\bar{I}).$$

From Lemma 2 it will then follow that $\lim_{z \to w} p(z) = 0$ for every $w \in I$ and hence for at least one $w \in K$, contrary to the definition of K. Thus Lemma 1 and, consequently, Theorem 2α will be proved. Write $I\backslash K$ as a countable union of open intervals and let (c, d) be one of these intervals.(If c and d are points of T, by (c, d) we will mean the shorter arc of T lying between c and d. Note $|c - d| = 2 \sin\frac{\theta}{2}$ when the arclength of (c, d) is θ.) By Lemma 3α below it follows that $p \leq M$ on T_{cd}. Since $S(\bar{I})$ is covered by the union of these triangles together with a union of regions $\Omega(w)$ where $w \in F_j$, (4) follows. QED

Lemma 3α. *Suppose p satisfies (1), (3), $p^*(c) \leq M$, $p^*(d) \leq M$, and $\lim_{z \to w} p(z) = 0$ for all w in the arc (c, d). Then $p(z) \leq M$ for all $z \in S([c, d])$.*

Proof. Let T_{cd} be the curvilinear triangle bounded by the arc (c, d) and the line segments pc and pd where p is the point of intersection of the straight side of $\Omega(c)$ closest to d and the straight side of $\Omega(d)$ closest to c. We need only show that $p(z) \leq M$ for $z \in T_{cd}$. Let F be a conformal mapping of T_{cd} onto the unit disc D. Then the angle at c of $\frac{\pi - \alpha}{2}$ is straightened out into an angle of π by F. In other words, neglecting a translation and a rotation, for $z \in T_{cd}$ near c, F is asymptotically $z^{\frac{2\pi}{\pi - \alpha}}$ (which maps e^{i0} to e^{i0} and $e^{i(\frac{\pi - \alpha}{2})}$ to $e^{i\pi}$). Hence if $|\zeta - F(c)| = \delta$, then $|F^{-1}(\zeta) - c| \simeq \delta^{\frac{\pi - \alpha}{2\pi}}$; so letting $v(\zeta) := p(F^{-1}(\zeta))$, it follows from (3) and Lemma 4 below that $|v(\zeta)| = o(\frac{1}{\delta})$. Similarly if $|\zeta - F(d)| = \delta$, then $v(\zeta) = o(\frac{1}{\delta})$. For $\eta \in (c, d)$, $\lim_{\zeta \to \eta} v(\zeta) = 0$, and for $\eta \in T\backslash[c, d]$, $v(\eta) = u(F^{-1}(\eta))$, where $F^{-1}(\eta) \in \Omega(c) \cup \Omega(d)$, so $v(\eta) \leq M$. Thus we may apply the Phragmén-Lindelöf Lemma 5 below to complete the proof of Lemma 3α. QED

Lemma 4. *Let $p(z)$ satisfy (1), $m(r) = o((1-r)^{-N})$ for some positive real number N, and $\lim_{z \to w} p(z) = 0$ for all w in the arc (c_1, c_2). Then, for both $i = 1$ and $i = 2$, as z tends to c_i, we have*

(5) $$p(z) = o(|z - c_i|^{-N}).$$

Proof. We observe that if B is a disk centered at z and v a function harmonic in B and continuous on $\bar{B}$ and if $p \in (0, 1)$, then

(6)
$$|v(z)| \leq C_p \left(\frac{1}{|B|} \iint_B |v(x, y)|^p \, dx dy \right)^{\frac{1}{p}}.$$

This result is stated and proved on pages 172–173 of [FS]. For our purposes we need to lighten the hypothesis for inequality (6) from harmonic to subharmonic and nonnegative.

This is straightforward, so we will limit ourselves to a few remarks on p. 173 of [FS]. In line 3, " $= r$" should be " $= r^2$"; in line 6, append "provided $p > 1 - \theta$"; and in line 11, change the second upper limit of integration to 1. The only real change involves establishing the estimate $m_\infty(\rho) \leq A(1 - \rho r^{-1})^{-n} m_1(r)$ on line 8 for a nonnegative subharmonic function p. The notation m_p is from [FS]. To see this, introduce the harmonic function h which agrees with p on the origin-centered spherical surface of radius r and note (i) the estimate holds for h, (ii) $m_\infty(\rho, p) \leq m_\infty(\rho, h)$ by the maximum principle, and (iii) $m_1(r, h) = m_1(r, p)$ by the definition of h.

Let $I := (c_1, c_2)$ and define

$$v(z) := \begin{cases} p(z), & \text{if } z \in S(I) \\ 0, & \text{if } z \notin S(I). \end{cases}$$

Then $v(z)$ is actually subharmonic and nonnegative on the infinite wedge $W := \{rw : w \in I, 0 \leq r < \infty\}$. This is easy to check: for example, if $w \in I$ and B is a ball about w small enough to be contained in W, then $v(w) = 0 \leq |B|^{-1} \int_{B \cap D} p = |B|^{-1} \int_B v$. We will deduce the required estimate (5) only at c_1, since the argument at c_2 is symmetric. Write $z = c_1 - \delta e^{i\varphi}$, so that the vector from c_1 to z makes an angle of φ, $0 < \varphi < \frac{\pi}{2}$, with the vector from c_1 to the origin. If $0 < \varphi \leq \frac{\pi}{4}$, then it is geometrically evident that there is an absolute constant C such that $|z - c_1|^{-1} \leq C(1 - |z|)^{-1}$, whence (5) is immediate. So assume $\frac{\pi}{4} < \varphi < \frac{\pi}{2}$, which insures that B, a disc of radius $\delta/8$ (say) about z is contained in W. Let A be the polar rectangle $\{(r, \theta) : 1 - 2\delta \leq r < 1, \arg(z) - \frac{\delta}{8} \leq \theta \leq \arg(z) + \frac{\delta}{8}\}$. Clearly $B \cap D \subset A$. Applying the inequality (6) we have, for any $q > 0$,

(7)
$$v(z)^q \leq C_q \frac{1}{|B|} \iint_B v(x, y)^q \, dx dy = C_q \frac{1}{|B|} \iint_{B \cap D} p(x, y)^q \, dx dy$$
$$\leq C_q \frac{1}{|B|} \int \int_A p(x, y)^q \, dx dy.$$

Now set $q := 1/2N$, change to polar coordinates, note $|B| = O(\delta^2)$, note that the hypothesis on $m(r)$ can be rewritten as $p(r, \theta) = o((1 - r)^{-N})$, and estimate the Jacobian r by 1. All this substituted into inequality (7), raised to the $2N$th power yields

$$v(z) = O\left(\delta^{-2} \int_{1-2\delta}^1 o((1 - r)^{-\frac{1}{2}}) \, dr \int_{\arg(z)-\delta/8}^{\arg(z)+\delta/8} d\theta \right)^{2N} = o(\delta^{-2+1/2+1})^{2N} = o(\delta^{-N}).$$

QED

Lemma 5. *(Phragmén-Lindelöf Lemma) Let $p(z)$ satisfy (1) and suppose that*

$$\limsup_{z \to w} p(z) \le M$$

for all $w \in T \backslash \{c_1, ..., c_n\}$. Suppose also that for each j, $1 \le j \le n$, we have

(8) $$|p(z)| = o(|z - c_j|^{-1}) \text{ as } z \in D \text{ tends to } c_j.$$

Then $p(z) \le M$ in D.

Proof. Let $D(c, \delta) := \{z : |z - c| < \delta\}$; $I(c, \delta) := D(c, \delta) \cap T$, so that I is the arc (c^-, c^+) of T where $\arg(c^\pm) = \arg(c) \pm \arcsin(\frac{\delta}{2})$; and ω^z be the harmonic measure for D so that if $E \subset T$, then $\omega^z(E)$ is the Poisson integral of the characteristic function of E. If $e^{i\theta_0} \in T$, $z = re^{i\theta} \in D$, and $|z - e^{i\theta_0}| = \delta$, we have the estimate

$$\omega^z(I(e^{i\theta_0}, \delta)) := \frac{1}{2\pi} \int_{c^-}^{c^+} \frac{1 - r^2}{1 - 2r \, \cos(\theta - \varphi) + r^2} d\varphi$$

$$\ge c \int_{\delta/4}^{\delta/4} \frac{1 - r}{(1 - r)^2 + (\theta - \theta_0 - \varphi)^2} d\varphi$$

$$= c \left\{ \arctan\left(\frac{\theta - \theta_0 + \delta/4}{1 - r} \right) - \arctan\left(\frac{\theta - \theta_0 - \delta/4}{1 - r} \right) \right\}.$$

Now $1 - r \le \delta$, so if $\theta \ge \theta_o$, the first arctan exceeds $\arctan(\frac{\delta/4}{\delta})$, while the second arctan is negative, so that $c \arctan(\frac{1}{4})$ is a lower bound. The case of $\theta \le \theta_o$ is symmetrical. In other words, there is a positive absolute constant c_o so that

(9) $$\omega^z(I(e^{i\theta_0}, \delta)) \ge c_o \text{ whenever } z \in D \text{ and } |z - e^{i\theta_0}| = \delta.$$

Fix $\delta > 0$; for each j, $1 \le j \le n$, let $u_j(z, \delta) := \omega^z(I(c_j, \delta))/c_o$; and define the domain $D_\delta := D \backslash (\cup_j D(c_j, \delta))$. Using the estimate (9) and applying the maximum principle for subharmonic functions in D_δ, we have

(10) $$p(z) \le M + \sum_{j=1}^{n} u_j(z, \delta) M_j(\delta)$$

for z in D_δ, where $M_j(\delta) := \sup\{p(z) : z \in \partial D(c_j, \delta) \cap D$. Now let $z = re^{i\theta} \in D$ and estimate as above to get

$$u_j(z, \delta) \le C \{ \arctan\left(\frac{\theta - \theta_j + \delta/4}{1 - r} \right) - \arctan\left(\frac{\theta - \theta_j - \delta/4}{1 - r} \right) \},$$

where θ_j is the argument of c_j.

Finally, freeze z and let $\delta \to 0$. We see that there is a constant $C_j(z)$, depending only on j and z, so that $u_j(z, \delta) \le C_j(z) \cdot \delta$ as $\delta \to 0$. Thus, taking hypothesis (8) into account, we see that $u_j(z, \delta) M_j(\delta) = O(\delta) o(\frac{1}{\delta}) = o(1)$ as $\delta \to 0$. Substituting these n relations into inequality (10) and letting $\delta \to 0$ establishes Lemma 5. QED

3. Examples

Theorem 4. *There is a function which is not identically 0, which is harmonic on D, and which has non-tangential limit 0 at every point of T.*

Proof. Let

$$f(z) := \int_0^\infty \frac{e^{zt}}{t^t} dt$$

Then f satisfies

(11)   there is a constant A so that $|f(z)| \leq A$ if $|Im(z)| \geq \pi$,

(12) f is entire and hence in particular continuous at each finite z, and

(13)   for x real, $f(x) \simeq \sqrt{2\pi}e^{\{e^{(x-1)}+\frac{1}{2}(x-1)\}}$.

(See [BN], pp.140-143, for properties (11) and (12). See the estimate following this proof for (13).) Let $S(z) := \pi i(\frac{1+z}{1-z}) - 2\pi i$. Direct calculation shows that $S((1-a)+ae^{i\theta}) = -\frac{\pi}{a}\cot\frac{\theta}{2} + (\frac{1}{a} - 3)\pi i$. Setting $a = 1$ shows that S maps $T = \partial D$ onto the line $L := \{x - 2\pi i : -\infty < x < \infty\}$. Similarly setting $a = \frac{1}{2}, \frac{1}{3}$, and $\frac{1}{4}$ shows that S maps each circle T_i onto the line L_i, $i = 1, 2, 3$, as shown,

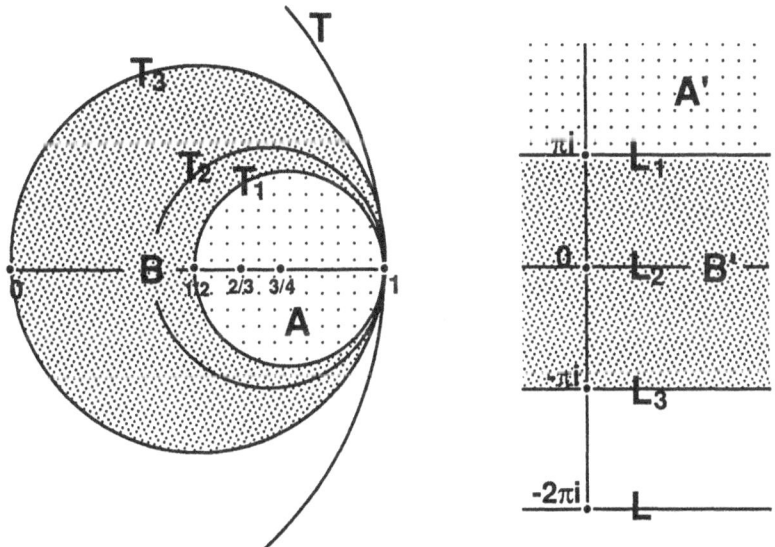

Also S maps the dotted disc A onto the dotted half plane A' and the dark shaded region B onto the similarly shaded strip B' as shown. Now let $g(z) := (1 - z)f(S(z))$. From (12) it follows that

(14) g is continuous at every point of T except $z = 1$.

Since (11) holds and $A' \subset \{z : |Im\ z| \geq \pi\}$, it follows that

(15) g has non-tangential limit 0 as z tends to 1.

Finally, let $h(z) := g(z) - PI(g|_T)(z)$, where $(PI)(f)$ denotes the Poisson integral of f. From (12) we see that $g|_T$ is continuous on $T\backslash\{1\}$. Motivated by (15), define $(g|_T)(1) := 0$. From (11) and $L \cap \{|Im(z)| < \pi\} = \emptyset$, we have

$$\lim_{\substack{w \in T \\ w \to 1}} g(w) = 0$$

so that $g|_T$ is continuous on all of T. Thus at every point w of T, the non-tangential limit as $z \to w$ of $PI(g|_T)(z)$ is $g(w)$.[Z], III.7.9 From (14) and (15) it follows that h has non-tangential limit 0 everywhere on T.

Now the Poisson kernel is positive and has integral 1 so that $\sup_{z\in D}|PI(g|_T)(z)| \leq \sup_{w\in T}|g(w)|$, which is finite since $g|_T$ is a continuous function on a compact set. To get $h \neq 0$ we will show g to be unbounded on D. On T_2 we have the estimates for $\theta > 0$ small,

$$g(\frac{2}{3} + \frac{1}{3}e^{-i\theta}) = (\frac{1}{3} - \frac{1}{3}e^{-i\theta})\int_0^\infty \frac{e^{(\frac{\pi}{2}\cot\frac{\theta}{2})t}}{t^t}dt$$

$$\simeq \frac{\sqrt{2\pi}}{3}i\theta e^{e^{\{\frac{\pi}{2}\cot\frac{\theta}{2}-1\}+\frac{1}{2}(\frac{\pi}{2}\cot\frac{\theta}{2}-1)}}$$

But if $d := \text{dist}\{\frac{2}{3} + \frac{1}{3}e^{-i\theta}, T\}$, $d = 1 - \frac{1}{3}\sqrt{5 + 4\cos\theta} \simeq (\frac{\theta}{3})^2$ so that $\theta \simeq 3\sqrt{d}$. Also, $\sqrt{2\pi} > \frac{5}{2}$ and $\frac{\pi}{2}\cot\frac{\theta}{2} \simeq \frac{\pi}{\theta}$, so for θ small, $\frac{\pi}{2}\cot\frac{\theta}{2} - 1 \approx \frac{\pi}{3\sqrt{d}} - 1 > \frac{1}{\sqrt{d}}$. Thus if $z = \frac{2}{3} + \frac{1}{3}e^{-i\theta}$ and if $d = 1 - |z|$, $\text{Im } g(z) > \frac{5}{2}\sqrt{d}e^{\{e^{1/\sqrt{d}}+1/(2\sqrt{d})\}}$.   QED

Remarks. In particular, if a real valued harmonic example is desired, $\text{Im } h$ will do. Note that $PI(g|_T)$ cannot be analytic even though g is analytic on D, for then h would be an analytic function with nontangential boundary values 0 on a set of positive measure, contrary to a well known theorem of Privalov.[Z], XIV.1.9

Estimate.

$$\int_0^\infty \frac{e^{xt}}{t^t}dt \simeq \sqrt{2\pi}e^{\{e^{x-1}+\frac{(x-1)}{2}\}} \qquad \text{as } x \to +\infty.$$

Proof. Let $t =: e^x s$. Then

$$I := \int_0^\infty (\frac{e^x}{t})^t dt = e^x \int_0^\infty (\frac{1}{s})^{e^x s}ds = e^x \int_0^\infty e^{e^x(s\log\frac{1}{s})}ds.$$

Let $u := s - e^{-1}$. Then

$$I = e^x \int_{-e^{-1}}^\infty e^{e^x[\ln(u)+e^{-1}]}du = e^x \cdot e^{e^{x-1}}\int_{-e^{-1}}^\infty e^{e^x h(u)}du$$

where $h(u) := (u + e^{-1})\ln\frac{1}{(u+e^{-1})} - e^{-1}$ is increasing on $[-e^{-1}, 0]$, zero at $u = 0$, and decreasing on $[0, \infty)$. Also $h''(0) = -e$, so using the estimate $h(u) = -e\frac{u^2}{2} + o(u^2)$ near 0 and the estimate $e^x h(u) < -(e^x - 1)(-h(d)) + h(u)$ for $|u| > d$ it is easy to show (See [De], pp. 63-65) that

$$\int_{-e^{-1}}^\infty e^{e^x h(u)}du \simeq \int_{-\infty}^\infty e^{(e^x)\frac{h''(0)}{2}u^2}du.$$

Since

$$\int_{-\infty}^{\infty} e^{-(e^x+1)\frac{u^2}{2}} du = \sqrt{\frac{2\pi}{e^x+1}},$$

$$I \simeq e^x e^{e^{x-1}} \sqrt{2\pi} e^{-\frac{1}{2}x-\frac{1}{2}} = \sqrt{2\pi} e^{\{e^{x-1}+\frac{1}{2}(x-1)\}} = \sqrt{2\pi e^{x-1}} e^{e^{x-1}}.$$

More precisely, ([De], pp. 66-69), $I = \sqrt{2\pi} e^{\{e^{x-1}+\frac{1}{2}(x-1)\}} + O\left(e^{\{e^{x-1}-\frac{1}{2}x\}}\right)$ or even

$$I = \sqrt{2\pi} \left[e^{\{e^{x-1}+\frac{1}{2}(x-1)\}} - \frac{1}{24} e^{\{e^{x-1}-\frac{1}{2}x-\frac{3}{2}\}} \right] + O(e^{\{e^{x-1}-\frac{3}{2}x\}}).$$

QED

Second proof of Theorem 4.(W. Rudin) For $\alpha > 0$, let Γ_α be the directed path in the complex plane obtained by traveling leftward along the ray from $(+\infty, -i\pi)$ to $(\alpha, -i\pi)$, then up the line segment from $(\alpha, -i\pi)$ to $(\alpha, +i\pi)$, and then rightward along the ray from $(\alpha, +i\pi)$ to $(+\infty, +i\pi)$. Let $P_\alpha := \{z : Re(z) < \alpha\}$. For $z \in P_\alpha$ define

$$f_\alpha(z) := \frac{1}{2\pi i} \int_{\Gamma_\alpha} \frac{e^{e^w}}{w-z} dw.$$

If $\alpha < \beta$, then, by Cauchy's theorem, $f_\alpha = f_\beta$ in P_α ; so there is an entire function f such that $f = f_\alpha$ in P_α. This f appears in exercise 11, chapter 16 of $[R]$ and has the following properties:

(16) $f(x)$ is real for $-\infty < x < \infty$,

(17) $f(re^{i\theta}) \to 0$ as $r \to \infty$, uniformly in $0 < \delta \le \theta \le 2\pi - \delta$, for all $\delta > 0$, and

(18) $f \not\equiv 0$ (because $f(x) = e^{e^x} + O(1)$ as $x \to +\infty$).

Now define $u(z) + iv(z) := f(i\frac{1+z}{1-z})$. Then v is harmonic at all $z \ne 1$. Relation (16) implies that $v(e^{i\theta}) = 0$ for all $e^{i\theta} \ne 1$, v has nontangential limit 0 at 1 (from inside D) because of (17), but $v \not\equiv 0$ because of (18). QED

References.

[BN] J. Bak and D.J. Newman, *Complex Analysis*, Springer, New York, 1982.

[D] B. E. J. Dahlberg, On the radial boundary values of subharmonic functions, *Math. Scand.* **40** (1977), 301-317.

[De] N.G. DeBruijn, *Asymptotic Methods in Analysis*, Dover, New York, 1981.

[FS] C. Fefferman and E. M. Stein, H^p spaces of several variables, *Acta Math.* **129** (1972), 137-192.

[R] W. Rudin, *Real and Complex Analysis*, McGraw-Hill, New York, 1987.

[S] V. Shapiro, The uniqueness of functions harmonic in the interior of the unit disk, *Proc. Lond. Math. Soc.* **13** (1963), 639-652.

[T] M. Tsuji, *Potential Theory in Modern Function Theory*, Maruzen, Tokyo, 1959.

[V] S. Verblunsky, On the theory of trigonometric series, (I) *Proc. Lond. Math. Soc.* **34** (1932), 441-456; (II) *ibid.*457-491.

[W] F. Wolf, The Poisson integral. A study in the uniqueness of harmonic functions, *Acta Math.* **74** (1941), 65-100.

[Z] A. Zygmund, *Trigonometric Series*, Vols.I, II, Cambridge Univ. Press, Cambridge, 1959.

ESTIMATES FOR THE KAKEYA MAXIMAL OPERATOR ON RADIAL FUNCTIONS IN $\mathbf{R}^n$

ANTHONY CARBERY, EUGENIO HERNÁNDEZ AND FERNANDO SORIA

University of Sussex, Universidad Autónoma de Madrid
and The Institute for Advanced Study, Princeton

Introduction.

For a real number $N > 1$, the Kakeya maximal operator K_N is defined on locally integrable functions f of $\mathbf{R}^n$ as

$$K_N f(x) = \sup_{x \in R \in \mathcal{B}_N} \frac{1}{|R|} \int_R |f(y)| dy,$$

where $\mathcal{B}_N$ denotes the class of all rectangles in $\mathbf{R}^n$ of eccentricity N, that is, congruent with any dilate of the rectangle $[0,1]^{n-1} \times [0,N]$, and where $|A|$ represents the Lebesgue measure of the set A.

It is conjectured that K_N is bounded on $L^n(\mathbf{R}^n)$ with a constant which grows no faster than $O((\log N)^{\alpha_n})$, for some $\alpha_n > 0$, as $N \to \infty$. This conjecture, which is closely related to a longstanding classical conjecture about the boundedness of Bochner-Riesz means in $\mathbf{R}^n$, was solved in the positive for the case $n = 2$ by Córdoba [2] (see also Fefferman [3], Strömberg [9] and Wainger [10]).

One of the purposes of this work is to present a simple proof of the conjecture in any dimension when K_N is restricted to the class of radial functions. The precise statement of this result is given in the following

Theorem A. *There exists a constant C_n such that for every radial function f one has*

(i) $$t|\{x \in \mathbf{R}^n : K_N f(x) > t\}|^{1/n} \leq C_n (1 + \log N)^{(n-1)/n} \|f\|_{L^n(\mathbf{R}^n)}$$

(ii) $$\|K_N f\|_{L^n(\mathbf{R}^n)} \leq C_n (1 + \log N) \|f\|_{L^n(\mathbf{R}^n)}.$$

It is not hard to see, moreover, that the dependence on N of the above constants is sharp by taking just $f(x) = \frac{1}{|x|} \chi_{1 \leq |x| \leq N}$. A similar result, but for a smaller Kakeya operator (averages only on rectangles congruent with $[0,1]^{n-1} \times [0,N]$) and with bigger exponents, was first obtained by Igari [6].

Given a finite set of unit vectors $\Omega = \{\omega_i\}_{i=1}^N$, we also define M_Ω as the maximal operator over rectangles in $\mathbf{R}^n$ having one side parallel to one of the given directions ω_i. When these N directions are uniformly distributed on S^{n-1}, we will simply write M_N instead of M_Ω. One can easyly see that in $\mathbf{R}^n$, K_N is majorized by $M_{N^{n-1}}$ and therefore Theorem A follows from the more general:

Theorem B. *Let Ω denote a collection of N unit vectors in $\mathbf{R}^n$. Then there exists a constant C_n such that*

(i') $$t|\{x \in \mathbf{R}^n : M_\Omega f(x) > t\}|^{1/n} \le C_n(1 + \log N)^{(n-1)/n}\|f\|_{L^n(\mathbf{R}^n)}$$

(ii') $$\|M_\Omega f\|_{L^n(\mathbf{R}^n)} \le C_n(1 + \log N)\|f\|_{L^n(\mathbf{R}^n)}$$

for every radial function f.

We would like to point out that, even for the case $n = 2$, it is still an open problem to determine whether a similar result for M_Ω holds on arbitrary functions. It turns out that for radial functions in $\mathbf{R}^n$, M_Ω is also of strong type p, for $p > n$, with a constant independent of Ω. This is quite different with respect to what happens in the general problem since it is known, from the construction of the Kakeya set, that the constant of boundedness in $L^p(\mathbf{R}^n)$ of K_N grows at least as $O((\log N)^{\alpha_{p,n}})$ for some $\alpha_{p,n} > 0$.

An even more interesting result holds for radial functions, namely, a certain type of boundedness at the end point, $p = n$, of the maximal operator given by averages over all rectangles in $\mathbf{R}^n$. This operator is of course not bounded in $L^p(\mathbf{R}^n)$, $n > 1$, except in the trivial case $p = \infty$ (see de Guzmán [5]). It is not of weak type n on radial functions either (see the remark following Lemma 2). We will show, however, that for radial functions it is of restricted weak type n (Proposition 4). The same is true then for K_N and for M_Ω. Theorems A and B will be obtained from this and a simple three point interpolation technique.

Theorems A and B were announced in [Tôhoku Math. Jour., vol.41 (1989), pp.647-656] but an error there lead to an incorrect proof. We are very grateful to Prof. S. Igari and the committee of the ICM-90 Satellite Conference on Harmonic Analysis, Sendai, for this opportunity to put matters straight.

Definitions and Notation. Given a locally integrable funcion f on $\mathbf{R}^n$ we denote by $\mathcal{M}f$ its associated "universal maximal function" (averages over all rectangles in $\mathbf{R}^n$). It is easy to see that $\mathcal{M}f$ is majorized by

$$\mathcal{M}_0 f(x) = \sup_{\omega \in S^{n-1}} \sup_{r>0} \frac{1}{r} \int_0^r |f(x + t\omega)|\,dt$$

(averages over all directions). Let us consider also the following one-sided maximal operators defined on functions $g : [0, \infty) \to \mathbf{C}$

$$Ag(t) = \sup_{0 < a < t} \frac{1}{(t^2 - a^2)^{1/2}} \int_a^t |g(s)| \frac{s}{(s^2 - a^2)^{1/2}} ds, \quad t > 0.$$

$$M_{\to} g(t) = \sup_{0 < t < b} \frac{1}{b - t} \int_t^b |g(s)| ds, \quad t > 0$$

And, similarly,

$$M_{\leftarrow} g(t) = \sup_{0 < a < t} \frac{1}{t - a} \int_a^t |g(s)| ds, \quad t > 0$$

Lemma 1. Let f be a radial function defined on $\mathbf{R}^n$ and let $f_0 : [0, \infty) \to \mathbf{C}$ be such that $f(x) = f_0(|x|)$. Then $\mathcal{M}_0 f(x) \leq C(M_{\leftarrow} f_0(|x|) + M_{\to} f_0(|x|) + A f_0(|x|))$, $\forall x \in \mathbf{R}^n$.

Remark. As it will be clear from the proof, one also has the inequality $A f_0(|x|) \leq \mathcal{M}_0 f(x)$. $M_{\leftarrow} f_0(|x|)$ and $M_{\to} f_0(|x|) \leq \mathcal{M}_0 f(x)$ is obvious from definition.

Proof of Lemma 1.

Since f is radial, $\mathcal{M}_0 f$ is also radial and therefore it suffices to take x of the form $x = (v, 0, 0, .., 0)$, $v > 0$. We will also assume, without loss of generality, f positive.

We fix $\omega = (\omega_1, \omega_2, ..., \omega_n) \in S^{n-1}$ and $r > 0$, and consider

$$I = \frac{1}{r} \int_0^r f(x + t\omega) dt = \frac{1}{r} \int_0^r f_0(\phi(t)) dt$$

where $\phi(t) = |x + t\omega| = (v^2 + 2t\omega_1 v + t^2)^{1/2}$.
    Observe that

$$\phi'(t) = \frac{\omega_1 v + t}{\phi(t)}$$

and that

$$\phi''(t) = \frac{v^2(1 - \omega_1^2)}{\phi^3(t)} \geq 0.$$

We shall divide the proof into four cases:

CASE 1. $\omega_1 \geq 0$.

In this case ϕ is increasing (on $(0,r)$) and by a change of variables we get

$$I = \frac{1}{r} \int_{\phi(0)}^{\phi(r)} f_0(u) W(u) du$$

where $W = (\phi^{-1}(u))' = \dfrac{1}{\phi'(\phi^{-1}(u))} = \dfrac{u}{\omega_1 v + \phi^{-1}(u)}$. Hence,

$$I = \frac{1}{r} \int_v^{\phi(r)} f_0(u) \frac{u}{\sqrt{u^2 - v^2(1 - \omega_1^2)}} du$$

Observe also that $W \geq 0$, W is decreasing on $(v, \phi(r))$ and

$$\frac{1}{r} \int_v^{\phi(r)} W(u) du = 1.$$

By an easy argument, similar to that in [8, Chap.III, Thm.2a], we can conclude

$$I \leq M_\rightarrow f_0(v).$$

CASE 2. $\omega_1 < 0, 0 < r < \dfrac{-\omega_1 v}{2}$.

In this case ϕ is decreasing and, therefore, so is ϕ^{-1} too. Changing variables again, we get

$$I = \frac{1}{r} \int_{\phi(r)}^{\phi(0)} f_0(u) \frac{u}{(-\omega_1 v) - \phi^{-1}(u)} du.$$

But, since $\phi(r) < u$ we have

$$0 < \phi^{-1}(u) < r < \frac{-\omega_1 v}{2}.$$

Thus,

$$I \leq \frac{2}{r} \int_{\phi(r)}^{\phi(0)} f_0(u) \frac{u}{-\omega_1 v} = \frac{-2}{r\omega_1 v} \int_{\phi(r)}^{v} f_0(u) u \, du.$$

We observe now that

$$\frac{-2}{r\omega_1 v} \int_{\phi(r)}^{v} u = \frac{-2\omega_1 v - r}{-\omega_1 v} \leq 2.$$

Arguing as above, we obtain here

$$I \leq 2M_\leftarrow f_0(v).$$

<u>CASE 3.</u> $\omega_1 < 0,\ \dfrac{-\omega_1 v}{2} \le r \le -2\omega_1 v.$

We can majorize I by

$$I \le \frac{2}{-\omega_1 v} \int_0^{-2\omega_1 v} f_0(\phi(t)).$$

and, since $\phi(t) = \phi(-2\omega_1 v - t)$, we see that

$$I \le \frac{4}{-\omega_1 v} \int_0^{-\omega_1 v} f_0(\phi(t))$$

$$= \frac{4}{-\omega_1 v} \int_{v(1-\omega_1^2)^{1/2}}^{v} f_0(u) \frac{u}{\sqrt{u^2 - v^2(1-\omega_1^2)}}.$$

If we now call $a = v(1 - \omega_1^2)^{1/2}$ we obtain, for $-1 < \omega_1 < 0$, that $0 < a < v$ and, therefore, in the present situation

$$I \le 4 \sup_{0 < a < v} \frac{1}{\sqrt{v^2 - a^2}} \int_a^v f_0(u) \frac{u}{\sqrt{u^2 - a^2}} du = 4A f_0(v).$$

<u>CASE 4.</u> $\omega_1 < 0,\ r > -2\omega_1 v.$

This case reduces to the previous ones by observing that

$$I \le \max\left(\frac{1}{-2\omega_1 v} \int_0^{-2\omega_1 v} f_0(\phi(t)),\ \frac{1}{r + 2\omega_1 v} \int_{-2\omega_1 v}^{r} f_0(\phi(t)) \right)$$

and that ϕ is increasing on $(-2\omega_1 v, \infty)$, with $\phi(-2\omega_1 v) = v$.

Therefore,

$$I \le \max(A f_0(v), M_\rightarrow f_0(v))$$

$$\#$$

<u>Lemma 2.</u> $\|M_\leftarrow g\|_{L^{n,\infty}(r^{n-1} dr)} \le \|g\|_{L^{n,1}(r^{n-1} dr)}.$ (With a similar result for $M_\rightarrow$.)

<u>Proof.</u> We may assume $g = \mathcal{X}_E$ with $\mathrm{E} \subseteq (0, \infty)$.

Let $\lambda > 0$ and consider $\{x > 0 \mid M_\leftarrow g(x) > \lambda\}$. If $\lambda > 1$ this is empty so we may assume $0 < \lambda \le 1$. It is a well known fact that we can write $\{x > 0 \mid M_\leftarrow g(x) > \lambda\} = \cup(a_k, b_k)$ with the $(a_k, b_k) \subset (0, \infty)$ pairwise disjoint and such that

$$\frac{1}{b_k - a_k} \int_{a_k}^{b_k} g \ge \lambda.$$

Let $d\mu_n(r) = r^{n-1}dr$ on $\mathbf{R}^+$; then

$$\mu_n(\cup(a_k,b_k)) = \sum_k \mu_n(a_k,b_k) = \frac{1}{n}\sum_k b_k^n - a_k^n.$$

Let $h_k = \int_{a_k}^{b_k} g$ so that $b_k \leq a_k + h_k/\lambda$. Using the elementary inequalities

$$(a+h/\lambda)^n - a^n \leq \frac{(a+h)^n - a^n}{\lambda^n}, \quad 0 < \lambda \leq 1,$$

(valid for $a, h > 0$ and $n \geq 1$) we see that

$$\mu_n(\cup(a_k,b_k)) \leq \frac{1}{n}\sum_k \frac{(a_k+h_k)^n - a_k^n}{\lambda^n} = \frac{1}{\lambda^n}\sum_k \int_{a_k}^{a_k+h_k} r^{n-1}dr$$

$$\leq \frac{1}{\lambda^n}\sum_k \int_{a_k}^{b_k} g(r)r^{n-1}dr \text{ (since } r^{n-1} \text{ is increasing)}$$

$$\leq \frac{1}{\lambda^n}\mu_n(E) \text{ (since the } (a_k,b_k) \text{ are disjoint).}$$

Therefore, we have

$$\mu_n\{M\_\mathcal{X}_E > \lambda\} \leq \frac{1}{\lambda^n}\mu_n(E).$$

#

Remark. Given $1 < p < \infty$, the same argument of Lemma 2 could be applied to show that the Hardy- Littlewood maximal operator maps $L^{p,1}(\mathbf{R}, w_p dx)$ into $L^{p,\infty}(\mathbf{R}, w_p dx)$, where $w_p(x) = |x|^{p-1}$. It is interesting to notice however, that w_p belongs to the Muckenhoupt class of weights A_q only when $q > p$. A similar boundedness behavior -restricted weak type p but not weak type p with respect to w_p- holds also for any standard Calderón- Zygmund singular integral operator.

Now, to estimate Ag we will look at the larger operator $\tilde{A}g$ defined below

$$\tilde{A}g(r) = \sup_{0<a<r\leq b} \frac{1}{(b^2-a^2)^{1/2}}\int_a^b \frac{g(u)udu}{(u^2-a^2)^{1/2}}$$

Lemma 3. $\|\tilde{A}g\|_{L^{n,\infty}(r^{n-1}dr)} \leq 2\|g\|_{L^{n,1}(r^{n-1}dr)}$, provided that $n \geq 2$.

Proof. As in the proof of Lemma 2, we let $g = \mathcal{X}_E$, so that $\tilde{A}g(r) \leq 1$. Hence for $0 < \lambda \leq 1$, using Young's interval selection lemma (see e.g. Garsia [4]), we may write $\{\tilde{A}g(x) > \lambda\} = \cup(a_k,b_k) \cup \cup(c_k,d_k)$ with the (a_k,b_k) pairwise disjoint and

$$\frac{1}{(b_k^2-a_k^2)^{1/2}}\int_{a_k}^{b_k} \frac{g(u)udu}{(u^2-a_k^2)^{1/2}} > \lambda.$$

(Similarly for the (c_k, d_k).)

Letting $h_k = \int_{a_k}^{b_k} \frac{g(u)u\,du}{(u^2 - a_k^2)^{1/2}}$, we see that $b_k^2 \leq a_k^2 + \frac{h_k^2}{\lambda^2}$. Hence,

$$\mu_n(\cup(a_k, b_k)) = \frac{1}{n} \sum_k (b_k^n - a_k^n)$$

$$\leq \frac{1}{n} \sum_k \left[\left(a_k^2 + \frac{h_k^2}{\lambda^2} \right)^{n/2} - (a_k^2)^{n/2} \right]$$

$$\leq \frac{1}{n} \frac{1}{\lambda^n} \sum_k \left[(a_k^2 + h_k^2)^{n/2} - a_k^{2n/2} \right]$$

(by the same elementary inequality as before)

$$= \frac{1}{\lambda^n} \sum_k \int_{a_k}^{(a_k^2 + h_k^2)^{1/2}} r^{n-1}\,dr.$$

Now if we set $|E \cap (a_k, b_k)| = \gamma_k$, then

$$h_k \leq \int_{a_k}^{a_k + \gamma_k} \frac{u\,du}{(u^2 - a^2)^{1/?}} = \{(a_k + \gamma_k)^2 - a_k^2\}^{1/2}$$

(Observe that the integrand is decreasing.) Thus, $(a_k^2 + h_k^2)^{1/2} \leq a_k + \gamma_k$, and so

$$\mu_n(\cup(a_k, b_k)) \leq \frac{1}{\lambda^n} \sum_k \int_{a_k}^{a_k + \gamma_k} r^{n-1}\,dr$$

$$\leq \frac{1}{\lambda^n} \sum_k \int_{a_k}^{b_k} g(r) r^{n-1}\,dr$$

(since r^{n-1} is increasing). Therefore,

$$\mu_n(\cup(a_k, b_k)) \leq \frac{1}{\lambda^n} \mu_n(E) \quad \text{and}$$

$$\mu_n(\tilde{A}\chi_E > \lambda) \leq \frac{2}{\lambda^n} \mu_n(E), \quad \text{by arguing similarly with} \quad (c_k, d_k).$$

#

Observation. Since $|\{x \in \mathbf{R}^n | f(x) > \lambda\}| = \mu_n\{t > 0 | f_0(t) > \lambda\}$ for a radial function f defined on $\mathbf{R}^n$, an operator mapping measurable radial functions to measurable radial functions will be bounded from (say) $L^{p,1}(\mathbf{R}^n) \to L^{p,\infty}(\mathbf{R}^n)$, when restricted to radial functions if and only if regarded as a 1-dimensional operator, it takes $L^{p,1}(r^{n-1}dr) \to L^{p,\infty}(r^{n-1}dr)$.

__Proposition 4.__ The universal maximal operator is of restricted weak type n on $\mathbf{R}^n$ when restricted to the class of radial functions.

__Proof.__ Combine Lemmas 1, 2 and 3.   #

To obtain other results for the Kakeya maximal operator or for the operator associated to N arbitrary directions (Theorems A and B) we need to notice that the universal maximal operator is bounded on L^∞ (with constant 1) and the maximal operator associated to N arbitrary directions, M_Ω, is of weak type q, for any $q > 1$, with constant $O(N)$.

__Proposition 5.__ Let $1 \le q < p$, and let T be a sublinear operator such that

$$T : L^q \to L^{q,\infty} \quad \text{with constant} \quad O(N)$$
$$T : L^\infty \to L^\infty \quad \text{with constant} \quad O(1)$$
$$T : L^{p,1} \to L^{p,\infty} \quad \text{with constant} \quad O(1)$$

$$\text{Then} \quad i) T : L^p \to L^{p,\infty} \quad \text{with constant} \quad O((\log N)^{1/p'})$$
$$ii) T : L^p \to L^p \quad \text{with constant} \quad O(\log N)$$

Moreover the same holds if in the hypothesis and conclusion we restrict T to the class of radial functions on $\mathbf{R}^n$.

First we will need the following

__Lemma 6.__ Given numbers $0 < a < b$, we have

$$\|f\chi_{a \le |f| \le b}\|_{p,1} \le C_p(1 + \log \frac{b}{a})^{1/p'}\|f\|_p$$

__Proof.__ Set $G = f\chi_{a \le |f| \le b}$ and let λ_G and λ_f denote the distribution functions of G and f, respectively.

Then

$$\|f\chi_{a \le |f| \le b}\|_{p,1} = \int_0^\infty \lambda_G^{1/p}(t)dt \le a\lambda_f^{1/p}(a) + \int_a^b \lambda_f^{1/p}(t)dt$$

Using Chebyshev's inequality in the first term and Hölder's in the second this expresion is bounded by

$$\|f\|_p + (\int_a^b \lambda_f(t)t^{p-1}dt)^{1/p}(\int_a^b \frac{1}{t}dt)^{1/p'}$$

#

Proof of Proposition 5.

We will assume, without loss of generality, that $\|Tf\|_\infty \leq \|f\|_\infty$. By slight abuse of notation, $|\ldots|$ will denote here the particular underlying measure being considered.

i) Fix $\lambda > 0$ and split $f = f\chi_{|f|\leq\lambda/3} + f\chi_{\lambda/3<|f|\leq\alpha\lambda} + f\chi_{|f|>\alpha\lambda}$ for α (large) to be chosen later. Thus, $f = f_1 + f_2 + f_3$, and $\{Tf > \lambda\} \subset \{Tf_1 > \lambda/3\} \cup \{Tf_2 > \lambda/3\} \cup \{Tf_3 > \lambda/3\}$. But $\{Tf_1 > \lambda/3\} = \emptyset$ and

$$|\{Tf_3 > \lambda/3\}| \leq C\frac{N^q\|f_3\|_q^q}{\lambda^q} \leq C\frac{N^q}{\alpha^{p-q}\lambda^p}\int_{|f|\geq\alpha\lambda}|f|^p.$$

On the other hand,

$$|\{Tf_2 > \lambda/3\}| \leq C\frac{\|f_2\|_{p,1}^p}{\lambda^p} = C\frac{\|f\chi_{\frac{\lambda}{3}\leq|f|\leq\alpha\lambda}\|_{p,1}^p}{\lambda^p}$$

$$\leq C(1+\log 3\alpha)^{p/p'}\|f\|_p^p/\lambda^p, \text{ by Lemma 6.}$$

Choosing $\alpha = N^{q/(p-q)}$ we see that $|\{Tf > \lambda\}| \leq C(1+\log N)^{p-1}\frac{\|f\|_p^p}{\lambda^p}$.

ii)

$$\int |Tf|^p = p\int_0^\infty \lambda^{p-1}|\{Tf > \lambda\}|d\lambda$$

$$\leq p\int_0^\infty \lambda^{p-1}\left[|\{Tf_2 > \lambda/3\}| + |\{Tf_3 > \lambda/3\}|\right]d\lambda.$$

Now,

$$\int_0^\infty \lambda^{p-1}|\{Tf_3 > \lambda/3\}|d\lambda \leq C\int_0^\infty \lambda^{p-1}\frac{N^q}{\lambda^q}\int_{|f|\geq\alpha\lambda}|f|^q dx d\lambda$$

$$= CN^q\int |f|^q\int_0^{|f|/\alpha}\lambda^{p-q-1}d\lambda dx = C\frac{N^q}{(p-q)\alpha^{p-q}}\int |f|^p dx;$$

and,

$$\int_0^\infty \lambda^{p-1}|\{Tf_2 > \lambda/3\}|d\lambda$$

$$\leq C(1+\log 3\alpha)^{p/p'}\int_0^\infty \lambda^{p-1}\frac{\|f_2\|_p^p}{\lambda^p}d\lambda \text{ (by the previous estimate)}$$

$$= C(1 + \log 3\alpha)^{p/p'} \int_0^\infty \int_{\frac{\lambda}{3} \leq |f| \leq \alpha\lambda} |f|^p dx \frac{d\lambda}{\lambda}$$

$$\leq C(1 + \log 3\alpha)^{p/p'} \int |f|^p \int_{|f|/\alpha}^{3|f|} \frac{d\lambda}{\lambda} dx$$

$$\leq C(1 + \log 3\alpha)^p \int |f|^p dx.$$

Choosing again $\alpha = N^{q/(p-q)}$ we see that

$$\int |Tf|^p \leq C(1 + \log N)^p \int |f|^p.$$

#

Remark. We would like to finish by mentioning that some of the ideas involved in this work have been further developed in [1] and [7] in connection with other interesting problems in Harmonic Analysis.

References

[1] A. Carbery, E. Romera and F. Soria, *Radial weights and mixed norm inequalities for the disc multiplier*, Preprint.

[2] A. Córdoba, *The Kakeya maximal function and the spherical summation multiplier*, Amer. J. Math 99 (1977), 1-22.

[3] C. Fefferman, *A note on the spherical summation multiplier*, Israel J. Math. 15 (1973), 44-52.

[4] A. Garsia, "Topics in almost everywhere convergence," Markham Publising Co., 1970.

[5] M. de Guzmán, "Differentiation of Integrals in $\mathbf{R}^n$," Springer-Verlag, 1975.

[6] S. Igari, *Kakeya's maximal function for radial functions*, preprint.

[7] P. Sjögren ans F. Soria, *Weak type (1,1) estimates for some extension operators related to rough maximal functions*, Preprint.

[8] E. M. Stein, "Singular Integrals and Differentiability properties of functions," Princeton U. Press, 1970.

[9] J. Strömberg, *Maximal functions associated to rectangles with uniformly distributed direccions*, Annals of Math. 107 (1978), 309-402.

[10] S. Wainger, *Applications of Fourier Transforms to averages over lower dimensional sets*, Proc. Symp. in Pure Math. **XXXV, part I** (1979), 85-94.

ADDRESSES. :

Anthony Carbery, *University of Sussex. Falmer, Brighton BN1-9QH*

Eugenio Hernández, *Universidad Autónoma de Madrid. 28049-Madrid*

Fernando Soria[1], *Univ. Autónoma de Madrid and The Institute for Advanced Study, Princeton, N.J. 08540*

[1]Partially supported by NSF grant DMS-8610730 and by DGITYC

SPECTRAL INVARIANTS OF CONFORMAL METRICS

SUN-YUNG A. CHANG AND PAUL C. YANG

1. Introduction. On a compact Riemannian manifold (M, g) the Laplace operator $\Delta = g^{-1/2} \partial_i (g^{1/2} g^{ij} \partial_j)$ acting on functions have discrete spectrum: $0 < \lambda_1 \leq \lambda_2 \leq \dots$ corresponding to the eigenfunctions

$$\Delta u + \lambda_i u = 0.$$

It is a well known problem of M. Kac to determine the geometry of (M, g) from the knowledge of the spectrum. There have been a number of examples of isospectral manifolds known since the 1960's. More recently the simple criteria for isospectrality of Sunada [Su] have lead to numerous constructions of isospectral metrics. In particular, it is even possible to construct continuous families of isospectral manifolds ([GW]).

In the known examples all the metrics appearing in a given isospectral set are in fact locally isometric. The global distinction arise from the way the isospectral pieces are glued together. It is therefore natural to suppose that this situation should be true for all isospectral manifolds.

As a preliminary step toward this direction it would be important to bound the size of any isospectral set of metrics. In [M] Melrose gave the first such compactness result for an isospectral set of plane domains. He showed that the heat invariants can be used to give bounds for the derivatives of the curvature functions of the boundary curves in an isospectral family of plane domains. This idea was extended by Brünning [Br] to give a similar compactness result for an isospectral set of potentials V for the Laplacian with potential $\Delta + V$ on a compact Riemannian manifold of dimension at most three.

Recently Osgood-Phillips-Sarnak considered the zeta function determinant of the Laplacian first introduced by Ray-Singer [RS]:

$$\log \det \Delta = -z'(0) \text{ where } z(s) = \sum_{k=1}^{\infty} \lambda_k^{-s}.$$

Polyakov has shown ([P]) that for a compact surface M, the zeta function determinant of a conformally related metric $g = e^{2w} g_0$ can be expressed in terms of the zeta function determinant of the original metric g_0 plus an integral involving the conformal factor w:

(1.1)
$$z'_g(0) - z'_{g_0}(0) = \frac{1}{3} \int_M |\nabla w|^2 + 2Kw$$

where K is the Gauss curvature of the background metric g_0. In [On] and [OPS1] it was shown that the standard metric of constant curvature realizes the extremum of the zeta function determinant among metrics of fixed area in a given conformal class. As a consequence, Wolpert's asymptotic calculation [W] for the zeta function determinant of degenerating hyperbolic metrics shows that the underlying conformal strucuture for an isospectral set of metrics is compact in the moduli space of conformal structures. The compactness question for isospectral metrics is thus reduced to that of an isospectral set of conformal metrics. Osgood-Phillips-Sarnak went on to show that an isospectral set of conformal factors is compact in the $\mathcal{C}^\infty$ topology again making use of the zeta function determinant. In this way they have shown the following

Theorem 1.1. [On] [OPS2] *Given an isospectral set of metrics g_α on a compact surface M, there exists a set of diffeomorphisms ϕ_α of M so that the pullback metrics $\phi_\alpha^* g_\alpha$ form a compact set in the $\mathcal{C}^\infty$ topology in the space of metrics on M.*

For plane domains, there is a result parallel to the case for closed surfaces:

Theorem 1.2. [OPS3] *An isospectral set of plane domains (with respect to either the Neumann or Dirichlet boundary condition) form a compact set in the $\mathcal{C}^\infty$ topology.*

Remarks
1. For conformal metrics on the 2-sphere, the determinant was used to provide the initial estimates for the Dirichlet integrals of the conformal factor. It is possible to substitute the knowledge of λ_1 for this purpose ([CY2, appendix]). For the case of Dirichlet problem on plane domains, the determinant is crucial for the initial estimates. While for the Neumann problem, it remains unclear whether λ_1 may be substituted in place of the determinant.
2. To control the conformal structure, Wolpert made use of the Selberg trace formula to relate asymptotically the length of the shortest geodesic with the determinant. In his thesis, Lundelius ([L]) gave an extension of Wolpert's result to finite volume hyperbolic surfaces .

In dimensions greater than two, the set of conformal structures on a given compact manifold is poorly understood. On the other hand there is a relatively good understanding of the set of metrics in a given conformal class. Thus we will restrict ourselves to the conformal metrics, although we remark that there are some recent results under restrictive assumptions on the sign of the curvature ([An], [BPP]).

It turns out that the spectral invariants considered also play a central role in much of the recent work on the Yamabe problem and the problem of prescribing scalar curvature. As a common feature of these problems the analysis centers around several sharp inequalities of Euclidean harmonic analysis. The extremal functions for the inequalities can concentrate and blow up in a geometrically controlled manner. A basic philosophy is to show that this is the the only source of non-compactness in these problems. In more picturesque language, one needs to show that bubbling does not occur except in the trivial way indicated above. In the following we will discuss a couple of recent developments in this direction.

It is a pleasure for us to thank Professors Igari and Arai for the invitation to the conference "Harmonic Analysis" in Sendai and to thank Professors Sunada, Kurakawa, and Fujii and for the invitation to the conference "Zeta function in geometry" in Tokyo.

2. Heat invariants and the Sobolev inequality.

For a compact Riemannian manifold (M, g_0) with scalar curvature R_0, the conformal Laplacian L_0 is defined by

$$(2.1) \qquad L_0 v = -c_n \Delta_0 v + R_0 v, \qquad \text{where} \quad c_n = \frac{4(n-1)}{n-2}.$$

Under a conformal change of metric $g = u^{4/(n-2)} g_0$ the scalar curvature R of g is given by the equation

$$L_0 u = R u^{N-1}, \qquad \text{where} \quad N = \frac{2n}{n-2}.$$

The conformal Laplacian L enjoys the following covariance property

$$(2.2) \qquad L v = u^{-(N-1)} L_0(uv).$$

It follows that the Sobolev quotient

$$Q[M] = \inf_v Q[v] \qquad \text{with} \quad Q[v] = \frac{\int c_n |\nabla v|^2 + R_0 v^2}{(\int v^N)^{2/N}}$$

is a conformal invariant. We will refer to the numerator in $Q[v]$ as the energy $E[v]$. It will be convenient to use the sign of $Q[M]$ to divide the set of Riemannian metrics into three classes. Thus a Riemannian manifold M is said to have a positive conformal class if $Q[M] > 0$, etc. The Euler equation for the functional $Q[v]$ is the equation (2.1) with R_0 given by $Q[M]$. The solution of the Yamabe problem ([Au1], [S]) involves two assertions about $Q[M]$. The first is the assertion that $Q[M] < Q[S^n]$ for all M not conformally equivalent to the standard S^n. This means in the case $Q[M]$ is positive that the following sharp form of the Sobolev inequality holds on M:

$$(2.3) \qquad Q[M](\int v^N)^{2/N} \leq \int c_n |\nabla v|^2 + R_0 v^2, \quad Q[M] < Q[S^n].$$

The second is the assertion that when $Q[M] < Q[S^n]$ the minimizing sequence for the Sobolev quotient converges to a positive solution of (2.1) with $R = Q[M]$. On S^n the minimum Sobolev quotient is achieved by conformal factors u of the form $u^N = |det d\phi|$ where ϕ is a conformal transformation of S^n. Such conformal factors can concentrate at a point, while on the rest of S^n the conformal factor tends to zero. In this case, it is possible to regain compactness by a consideration introduced by Aubin ([Au2]). Given any conformal factor u on S^n there exist a conformal transformation ϕ of S^n which moves the conformal factor to a balanced position: setting $u_\phi^N = (u \circ \phi)^N |det d\phi|$ we have

$$\int u_\phi^N x_j = 0, \qquad 1 \leq j \leq n+1.$$

We shall denote by S the set of centered conformal factors. For conformal factors $u \in S$ there is a sharpened version of (2.2): there exists a constant $a = a(n) < 1$ so that for $u \in S$

$$(2.4) \qquad Q[S^n](\int u^N)^{2/N} \leq ac_n \int |\nabla u|^2 + R_0 \int u^2.$$

The above sharpened inequality provides the basic analytic tool for our approach to the problem of prescribing scalar curvature on S^n ([CY5]). The analytic difficulty here is the presence of the large conformal group. Given any conformal factor u on the standard n-sphere (S^n, g_0), we may move the conformal factor by a conformal transformation ϕ of the sphere: $u_\phi^{\frac{4}{n-2}} g_0 = \phi^* u^{\frac{4}{n-2}} g_0$. Since the energy is preserved: $E[u] = E[u_\phi]$, it is not possible to bound the conformal factors by the energy alone. In fact for very large conformal transformation ϕ the L^2 term in $E[\phi]$ tends to zero. This sharpened inequality allow us to give a lower bound for the L^2 norm of balanced conformal factors.

The Sobolev quotient is expressible in terms of the heat invariants associated to the Laplacian of the conformal metric:

$$tr\, e^{-t\Delta} \sim \sum_{k=0}^{\infty} a_k t^{-\frac{n}{2}+k} \quad \text{as } t \to 0.$$

The first few heat invariants are ([MS]):

$$a_0 = (4\pi)^{-n/2} \int dV$$

$$a_1 = (4\pi)^{-n/2} \frac{1}{6} \int R dV$$

$$a_2 = (4\pi)^{-n/2} \frac{1}{360} \int [5R^2 - 2\sum |R_{ij}|^2 + \sum |R_{ijkl}|^2] dV.$$

The Sobolev quotient of the positive conformal factor u is thus given as $\frac{6a_1(u^{4/(n-2)}g_0)}{a_0(u^{4/(n-2)}g_0)}$. In case n=3 the integrand in a_2 can be simplified to read:

$$a_2 = (4\pi)^{-n/2} \frac{1}{360} \int [3R^2 + 6\sum |R_{ij}|^2] dV.$$

In case n=4 we can extract a multiple of the Gauss-Bonnet integrand from a_2 to find:

$$a_2 = (4\pi)^{-n/2} \frac{1}{360} \int [3R^2 + 6\sum |R_{ij}|^2] dV + \frac{64\pi^2}{360} \chi(M).$$

For the higher order heat invariants, Gilkey ([G]) gave the coefficients of the leading order terms:

$$a_k = (-1)^k \int A_k |\nabla^{k-2} R|^2 + B_k |\nabla^{k-2} Ric|^2 + E_k dV,$$

where $A_k, B_k > 0$ and E_k is a polynomial of weight $2k$ in the derivatives of the curvature tensor of order at most $k - 3$.

Thus knowledge of a_2 gives L^2 bound for the Ricci tensor. It is possible to extend the compactness result to the conformal metrics in dimension three:

Theorem 2.1. [CY1] [CY2] *On a compact 3-manifold* (M, g_0), *given a conformal metric* $g = u^4 g_0$ *satisfying the conditions*
(α) $\int u^6 dV_0 = \int dV = 1$,
(β) $\int \sum |R_{ij}|^2 dV \leq C_1$,
(γ) $\lambda_1(g) \geq \Lambda > 0$,
then in case M *is conformally different from the standard* S^3, *there exists constant* $C = C(C_1, \Lambda)$ *so that*

(2.5)
$$\frac{1}{C} \leq u \leq C, \quad \text{and} \quad ||u||_{2,2} \leq C.$$

In case $M = S^3$, *we can replace condition* (β) *by*
(β') $\int |R|^2 dV \leq C_1$,
and assert that a suitable conformal transformation ϕ *exists so that the conclusion in* (2.5) *holds for* u_ϕ *where* $u_\phi^4 g_0 = \phi^* u^4 g_0$.

Remarks
1. This result gives an analytic characterization of the standard 3-sphere among all conformal classes of 3-manifolds. As a consequence we have another proof in the 3-dimensional case of a well known result of Obata that a compact 3-dimensional Riemannian manifold with non-compact group of conformal transformation is conformally equivalent to the standard 3-sphere.
2. In the case (M, g_0) has a negative conformal class the same conclusion hold with the condition (β') in place of (β) [BPY].
3. Since the conditions of the theorem involve only spectral data, we have as a corollary that an isospectral set of conformal factors on a compact 3-manifold form a compact set in the C^∞ topology. To obtain bounds for the higher order derivatives, we made use of the ellipticity of the scalar curvature equation and the calculation of Gilkey [G] for the coefficients of the leading order terms of the heat invariants a_k.
4. The first eigenvalue of the Laplacian plays the same role here as the determinant of the Laplacian does in the case of compact surfaces to prevent the formation of a thin neck.
    Recently Gursky ([Gu]) in his forthcoming thesis has extended this result in the following way:

Theorem 2.2. [Gu] *On a compact manifold* (M, g_0) *with Sobolev quotient* $Q[M] \neq 0$, *given a conformal metric* $g = u^{4/(n-2)} g_0$ *satisfying the conditions:*
(α) $\int u^N dV_0 = \int dV = 1$,
(β) $\int |\sum R_{ij}|^p dV \leq C_1$, *where* $p > \frac{n}{2}$,
then in case M *is not conformally the standard* S^n, *there exist constant C so that*

(2.6)
$$\frac{1}{C} \leq u \leq C, \quad \text{and} \quad ||u||_{2,p} \leq C;$$

in case $M = S^n$, *a suitable conformal transformation* ϕ *exists so that the conclusion in* (2.6) *holds for* u_ϕ *where* $u_\phi^{4/(n-2)} g_0 = \phi * u^{4/(n-2)} g_0$.

Remarks.
1. This result removes the assumption in Theorem 2.1 on the first eigenvalue $\lambda_1(g)$, and

represents a sharpening of the 3-dimensional result in the case M has a positive conformal class. In the case of the 3-sphere the conditions are different and complementary. There are examples showing these conditions are optimal.

2. This result again exhibits an analytic characterization of the standard n-sphere from other conformal classes modulo the case of zero scalar curvature.

3. This result can be applied to show C^∞ compactness of conformal factors with isospectral conformal Laplacians on a compact 3-manifold. The lowest eigenvalue of the conformal Laplacian does not yield as much information as the first eigenvalue of the Laplacian. The assumption of a bound for the L^p norm of the Ricci tensor effectively prevents the formation of a thin neck.

3. Adam's inequality and zeta function determinants in 4-dimension.

In dimension two the zeta functon determinant was central to the analysis of metrics on account of the formula of Polyakov relating the determinants of two conformally related metrics g_0 and $g = e^{2w}g_0$. The existence of such a formula came about because the two Laplace operators are closely related: $\Delta = e^{-2w}\Delta_0$. In general a geometrically defined operator A is said to have conformal covariance if

$$(3.1) \qquad A(\psi) = e^{-bw}A_0(e^{aw}\psi)$$

for some constants a and b. The well known examples of such operators are the conformal Laplacian (in this section it is more convenient to write the conformal factor in exponential form, thus we have $u^{\frac{4}{n-2}} = e^{2w}$):

$$Lv = -c_n\Delta v + Rv, \quad \text{where} \quad c_n = \frac{4(n-1)}{n-2}.$$

and the Dirac operator $\slashed{\nabla}$ on spinors where $\slashed{\nabla} = \gamma^i\nabla_i$ where $\gamma^1, \gamma^2, ...\gamma^n$ form an orthonormal frame and act on the spinors as elements of the Clifford algebra acting on the Clifford module of spinors.

The covariance property for L is given in (2.3) and the covariance property for $\slashed{\nabla}$ is given by

$$g = e^{2w}g_0, \quad \gamma = e^{-w}\gamma_0 \Rightarrow \slashed{\nabla} = e^{-(n+1)w/2}\slashed{\nabla}_0 e^{(n-1)w/2}.$$

For general elliptic operators A the zeta function determinant is defined ([RS]) as

$$-logdet(A) = \zeta_A'(0) \ where \ \zeta(s) = \sum_{\lambda_j\neq 0} \lambda_j^{-s}$$

$$= \frac{1}{\Gamma(s)}\int_0^\infty tr\ e^{-tA}t^{s-1}dt.$$

When one joins the conformally related metrics $g = e^{2w}g_0$ to g_0 by the curve of conformal metrics $g_t = e^{2tw}g_0$, it is possible to calculate the variation of the operators A_t and perform the integration in t to obtain analogues of the Polyakov formula (see [BO2]). As a special case we shall consider the conformal Laplacian L associated to conformal metrics on the 4-sphere. In this case the Branson-Ørsted formula reads as: denoting by v_0 the volume of the standard 4-sphere,

$$\frac{1}{4}(-log det(L) + log det(L_0)) = -32\pi^2 \frac{\int w}{v_0} - \int (\Delta w)^2 + 6 \int |dw|^2 - 2 \int (\frac{\Delta e^w}{e^w})^2.$$

Using this formula we can derive $W^{2,2}$ bounds for conformal factors with given determinants and volume:

Theorem 3.1. [BCY] *Suppose* $g = e^{2w} g_0$ *is a conformal metric on the standard 4-sphere satisfying*
$(\alpha) \int e^{4w} = volume(g) = c_1$,
$(\beta) log det(L) \leq c_2$,
then there exist constant $C = C(c_1, c_2)$ *and a conformal transformation* ϕ *of* S^4 *so that the transformed conformal factor* $e^{2w_\phi} g_0 = \phi^* e^{2w} g_0$ *satisfies* $\|w_\phi\|_{2,2} \leq C$.

Remarks.
1. In contrast to the situation in the previous section the heat invariant a_2 for the conformal Laplacian on 4-manifold is a conformal invariant. In general even dimensions $a_{n/2}$ is a conformal invariant for the conformal Laplacian ([BO1], [PR])
2. This result in fact holds for all compact Einstein locally symmetric spaces in 4-dimension except the hyperbolic spaces. This list consists of $S^4, CP^2, S^2 \times S^2$, 4-torus with their canonical metrics, and their quotients and the compact quotients of the complex polydisc and the complex 2-ball with their Bergman metrics. In fact the result remains valid for metrics which are sufficiently close to those metrics listed except for the standard 4-sphere where the situation is not clear.
3. The result continues to hold for the operator Dirac squared ∇^2 as well whenever spinor bundles exists on such spaces.
4. As a consequence of the $W^{2,2}$ bound it is in fact possible to give bounds for the L^2 norm of the Ricci tensor. Thus on such 4-manifold determinant is a strong substitute for a_2.

The analysis of the determinant rests upon an analogue of a sharp version of the Moser-Trudinger inequality found by Adams ([Ad]):
There exists a constant C such that for functions $w \in C_0^2(R^4)$ satisfying $\int (\Delta w)^2 \leq 1$ we have:

$$\int e^{32\pi^2 w^2} \leq C.$$

This inequality has an analogue for all compact 4-manifold:
There exists a constant $C = C(M, g)$so that for functions $w \in C^2(M)$ satisfying $\int w = 0$ we have

$$\int e^{32\pi^2 w^2} \leq C.$$

4. Extremals of the determinant.

A special case of Polyakov's formula (1.1) for S^2 with $g = e^{2w}g_0$ takes the form

$$z_g'(0) - z_{g_0}'(0) = \frac{1}{3}S[w] \quad where \quad S[w] = \frac{1}{4\pi}\int[|\nabla w|^2 + 2w], \quad if \int e^{2w} = 4\pi.$$

In [On] [OPS1] the extremal of the determinant of the 2-sphere is characterized:

Theorem 4.1. *For all conformal metrics* $g = e^{2w}g_0$ *of area* 4π *on* S^2, *we have* $S[w] \geq 0$ *with equality holding only for* w *of the form* $e^{2w}g_0 = \phi^*g$, *where* ϕ *is a conformal transformation of* S^2.

As in the case of Sobolev inequality, when we restrict ourselves to balanced conformal factors on S^2 satisfying

$$\int e^{2w}x_j = 0, \quad for \ j = 1, 2, 3,$$

we have a sharpened inequality: there exist a constant $a < 1$ so that

(4.1) $$\log\frac{1}{4\pi}\int e^{2w} \leq \frac{1}{4\pi}[\int a|\nabla w|^2 + 2w].$$

This inequality has played a central role in our work [CY3] [CY4] on the problem of prescribing scalar curvature on the 2-sphere. As we indicated before, the presence of the large conformal group makes it not possible to bound the Dirichlet integral $\int |\nabla w|^2$ by $S[w]$. However by restriction ourselves to centered conformal factors, the sharpened inequality allows us to split the terms in $S[w]$ to recover such a desired bound.

For conformal metrics on the 4-sphere, Theorem 3.1 allows us to take limit in a minimizing sequence to obtain

Theorem 4.2. [BCY] *On the standard 4-sphere given any conformal metric* $g = e^{2w}g_0$, *we have*

$$logdet(L) \leq logdet(L_0), \quad logdet(\nabla^2) \leq logdet(\nabla_0^2).$$

In each case equality holds if and only if $e^w = |d\phi|$, *where* ϕ *is a conformal transformation of the 4-sphere.*

For the proof of the theorem above, we require in addition to the Adams inequality another sharp inequality discovered by Beckner ([Be]): If $f \in C^\infty(S^n)$ has an expansion $\sum_{k=0}^\infty Y_k$ in spherical harmonics, then

(4.2) $$\log\int e^{f-\bar{f}} \leq \frac{1}{2n}\sum_{k=1}^\infty B(n,k)\int|Y_k|^2,$$

where $B(n,k) = \Gamma(n+k)/(\Gamma(n)\Gamma(k))$. Equality holds if and only if $e^{2f/n}g_0 = \phi^*g_0$ for some conformal transformation ϕ of the standard sphere (S^n, g_0).

In view of Beckner's inequality, which is valid on the standard n-sphere for any n, it is natural to speculate that Theorem 4.2 should be true for all even n.

REFERENCES

[Ad] D. R. Adams, *A sharp inequality of J. Moser for higher order derivatives*, Annals of Math. **128** (1988), 385–398.

[An] M. Anderson, *Remarks on the compactness of isospectral sets in low dimensions*, preprint.

[Au1] T. Aubin, *Equations différentielles non linéaire st problème de Yamabe concernant la courbure scalaire*, J. Math. Pures Appli. **55**, 269-296.

[Au2] T. Aubin, *Meilleures constantes dans le théorèm d'inclusion de Sobolev et un théorèm de Fredholm nonlinéaire pour la transformation conforme de la courbure scalaire*, J. Funct. Anal. **32** (1979), 148–174.

[Be] W. Beckner, *Sharp Sobolev inequalities on the sphere and the Moser-Trudinger inequality*, preprint.

[BCY] T. Branson, S.-Y. A. Chang and P. Yang, *Estimates and extremals for zeta function determinants in 4-manifolds*, preprint.

[BO1] T. Branson and B. Ørsted, *Conformal indices of Riemannian manifolds*, Compositio Math. **60**, 261-293.

[BO2] T. Branson and B. Ørsted, *Explicit functional determinants in four dimensions*, to appear, Proc. Amer. Math. Soc.

[BPP] R. Brooks, P. Perry, and P. Petersen, *Compactness and finiteness theorems for isospectral manifolds*, preprint.

[BPY] R. Brooks, P. Perry, and P. Yang, *Isospectral sets of conformally equivalent metrics*, Duke Jour. Math. **58**, 131-150.

[Br] J. Bruning, *On the compactness of isospectral potentials*, Comm. in Part. Diff. Equations 9, 687-698.

[CY1] S.-Y. A. Chang and P. Yang, *Compactness of isospectral conformal metrics on S^3*, Comment. Math. Helvetici **64** (1989), 363–374.

[[CY2] S.-Y. A. Chang and P. Yang, *Isospectral onformal metrics on 3-manifolds*, Jour. Amer. Math. Society **3**, 117-145.

[CY3] S.-Y. A. Chang and P. Yang, *Prescribing Gaussian curvature on S^2*, Acta Math. **159**, 215-259.

[CY4] S.-Y. A. Chang and P. Yang, *Conformal deformation of metrics on S^2*, Jour. Differential Geometry **27**, 259-296.

[CY5] S.-Y. A. Chang and P. Yang, *A perturbation result in prescribing scalar curvature on S^n*, preprint.

[G] P. Gilkey, *Leading terms in the asymptotics of the heat equation*, Contemporary Math. **73**, 79-85.

[GW] C. Gordon and E. Wilson, *Isospectral deformations of compact solvmanifolds*, Jour. Differential Geometry **19**, 241-256.

[Gu] M. Gursky, *Cal Tech Thesis*, in preparation.

[L] R. Lundelius, *Stanford thesis*.

[MS] H. McKean and I. Singer, *Curvature and the eigenvalues of the Laplacian*, Jour. Diff. Geometry 1, 43-69.

[M] R. Melrose, *Isospectral sets of drumheads are compact in C^∞*, preprint.

[On] E. Onofri, *On the positivity of the effective action in a theory of random surfaces*, Commun. Math. Phys. **86**, 321–326.

[OPS1] B. Osgood, R. Phillips, and P. Sarnak, *Extremals of determinants of Laplacians*, J. Funct. Anal. **80** (1988), 148–211.

[OPS2] B. Osgood, R. Phillips, and P. Sarnak, *Compact isospectral sets of surfaces*, J. Funct. Anal. **80** (1988), 212–234.

[OPS3] B. Osgood, R. Phillips, and P. Sarnak, *Moduli space, heights, and isospectral set of plane domains*, Annals of Math. **129**, 293-362.

[PR] T. Parker and S. Rosenberg, *Invariants of conformal Laplacians*, Jour. Diff. Geometry **25**, 199-222.

[P] A. Polyakov, *Quantum geometry of Bosonic strings*, Phys. Lett. B **103** (1981), 207-210.

[RS] D. Ray and I. Singer, *R-torsion and the Laplacian on Riemannian manifolds*, Advances in Math. **7**, 145-210.

[S] R. Schoen, *Conformal deformation of a Riemannian metric to constant scalar curvature*, Jour. Diff. Geometry **21**, 479-495.

[Su] T. Sunada, *Riemannian coverings and isospectral manifolds*, Annals of Math. **121**, 169-186.

[W] S. Wolpert, *Asymptotics of the spectrum and the Selberg zeta function on the space of Riemann surfaces*, Comm. Math. Physics **112**, 283-315.

Address:
Sun-Yung A.Chang, Department of Mathematics, UCLA, Los Angeles, CA 90024
Paul Yang, Department of Mathematics, USC, Los Angeles, CA 90089

REMARKS ON
THE BREAKDOWN OF ANALYTICITY
FOR $\bar{\partial}_b$ AND SZEGÖ KERNELS

MICHAEL CHRIST

UCLA

This expository article represents an expanded account of a lecture delivered at the conference on harmonic analysis held in Sendai in August, 1990. We aim primarily to present background material and to link together several recent works, but some new material is included as well.

Our principal theme is that certain very natural differential equations admit solutions which are not real analytic. Of prime concern for us is the operator $\bar{\partial}_b$, which arises in the analysis of boundary values of holomorphic functions on domains in $\mathbb{C}^2$, and the associated second-order operator $\bar{\partial}_b \bar{\partial}_b^*$. Another object of interest in complex analysis is the Szegö kernel, the distribution-kernel for the operator defined by orthogonal projection of L^2 onto the space of boundary values of holomorphic functions, on the boundary of a smoothly bounded domain in $\mathbb{C}^2$. The Szegö kernel is C^∞ off of the diagonal for a natural class of domains, but fails to be real analytic in certain simple examples.

1. Review of analytic hypoellipticity.

By "analytic" we shall always mean real analytic, and we denote by C^ω the class of analytic functions on some domain. All differential operators considered are assumed to have analytic coefficients. Recall that f is analytic on an open set U if and only if on each compact subset of U, there exists $C < \infty$ such that at every point,

$$|\partial^\alpha f| \le C^{|\alpha|+1} \alpha! \qquad \forall \alpha.$$

More generally, the Gevrey classes G^s are defined by the inequalities

$$|\partial^\alpha f| \le C^{|\alpha|+1} \Gamma(s|\alpha|) \qquad \forall \alpha.$$

Research supported in part by a grant from the National Science Foundation.

Definition 1.1. *A differential operator L is said to be analytic hypoelliptic at ζ if whenever u is a distribution and $Lu \in C^\omega$ in some neighborhood of ζ, then $u \in C^\omega$ in some neighborhood of ζ.*

Some relevant facts:

(1) Elliptic operators with analytic coefficients are analytic hypoelliptic [H, p. 306].
(2) No non-elliptic operator with constant coefficients is analytic hypoelliptic.
(3) Subelliptic operators of principal type are analytic hypoelliptic; see for instance [Tv1],[Tp]. Principal type means that where the principal symbol $p(x, \xi)$ vanishes, $dp(x, \xi)$ is not proportional to the one-form $\xi \cdot dx$; we will be concerned with operators whose principal symbol vanishes to order two, hence are not of principal type.
(4) Analyticity is a microlocal property. Oversimplifying a bit, f is said to be analytic in a conic subset Γ of $T^*\mathbb{R}^n$ if there exists $\varepsilon > 0$ such that $|\hat{f}(\xi)| = O(e^{-\varepsilon|\xi|})$ as $|\xi| \to \infty$ in Γ; but see [H],[Sj] for the correct formulation. Then a function is analytic in a neighborhood of x if it is microlocally analytic in several open conic subsets of phase space, the union of whose projections onto $\mathbb{R}^n$ contains x.
(5) Consider the following two examples:

$$\frac{\partial^2}{\partial x_1^2} + x_1^2 \frac{\partial^2}{\partial x_2^2} \qquad \text{in } \mathbb{R}^2,$$

$$\frac{\partial^2}{\partial x_1^2} + x_1^2 \frac{\partial^2}{\partial x_2^2} + \frac{\partial^2}{\partial x_3^2} \qquad \text{in } \mathbb{R}^3.$$

While the former is analytic hypoelliptic [Tv2], the latter is not [BG],[H, p. 310].

2. $\bar{\partial}_b$ and conjectures.

Rather than introducing three-dimensional CR manifolds and $\bar{\partial}_b$ in full generality, we shall restrict attention to an important class of examples. For various issues these examples serve as models for the general case [FK],[C1],[Sm], as the Heisenberg group does for the strictly pseudoconvex case [FS]. Consider the hypersurface $M = \{z \in \mathbb{C}^2 : \Im(z_2) = P(z_1)\}$, where P is a (real) analytic function defined on some subset of $\mathbb{C}$. We assume always that $\triangle P \geq 0$, which means that the domain above M is pseudoconvex, and that $\triangle P$ does not vanish identically, which is to say that M is of finite type.

The complex vector field

$$\bar{\partial}_b = \frac{\partial}{\partial \bar{z}_1} - i \frac{\partial P}{\partial \bar{z}_1} \frac{\partial}{\partial \bar{z}_2}$$

is tangent to M, and annihilates restrictions of holomorphic functions. Identify M with

$\mathbb{C} \times \mathbb{R}$ via the map $\mathbb{C} \times \mathbb{R} \ni (z,t) \mapsto (z, t + iP(z))$. Then $\bar{\partial}_b$ pulls back to

$$\bar{\partial}_b = \frac{\partial}{\partial \bar{z}} - i\frac{\partial P}{\partial \bar{z}}\frac{\partial}{\partial t}.$$

We normalize $\partial/\partial \bar{z}$ to be $\partial_x + i\partial_y$, and write more simply $\bar{\partial} = \partial/\partial \bar{z}$. We may write $\bar{\partial}_b = X + iY$ where

$$X = \partial_x + (\partial_y P)\partial_t, \qquad Y = \partial_y - (\partial_x P)\partial_t.$$

Set $\bar{\partial}_b^* = -X + iY$, the formal adjoint of $\bar{\partial}_b$ with respect to Lebesgue measure on $\mathbb{C} \times \mathbb{R}$. The following regularity properties of $\bar{\partial}_b$ are well known:

(1) $\bar{\partial}_b$ is not subelliptic nor even C^∞ hypoelliptic: L^2 solutions to $\bar{\partial}_b u \equiv 0$ need not be C^∞. Indeed, at any strongly pseudoconvex point w of M, there exists a linear holomorphic function h such that $h(w) = 0$ and $\Re(h) > 0$ at every other point of M in a neighborhood of w. If $\varepsilon > 0$ is small enough, then $h^{-\varepsilon}$ will be in L^2. It is annihilated by $\bar{\partial}_b$, and is unbounded.

(2) If $\bar{\partial}_b \bar{\partial}_b^* u \in C^\infty(U)$, then $\bar{\partial}_b^* u \in C^\infty(U)$; see [K], where it is also explained why this sort of regularity is natural to study.

This motivates the

Definition 2.1. $\bar{\partial}_b$ is relatively analytic hypoelliptic at ζ if whenever $u \in L^2$ and $\bar{\partial}_b \bar{\partial}_b^* u \in C^\omega$ in some neighborhood of ζ, then also $\bar{\partial}_b^* u \in C^\omega$ in some neighborhood.

Henceforth we abbreviate 'analytic hypoelliptic' and 'relatively analytic hypoelliptic' as AH and RAH, respectively. More generally, one has the analogous notion of relative G^s hypoellipticity, concerning regularity in the Gevrey classes. Assume that $\nabla P(0) = 0$. Let $\xi \in \mathbb{R}^2$, $\tau \in \mathbb{R}$ be coordinates dual to z, t. Let

$$\begin{aligned} U &= \{|(z,t)| < \varepsilon\}, \\ \Gamma^+ &= \{(z,t,\xi,\tau) : (z,t) \in U, \ \tau > 0 \text{ and } \tau \geq |\xi|\}, \\ \Gamma^- &= \{(z,t) \in U, \ \tau < 0 \text{ and } -\tau \geq |\xi|\}, \\ \Gamma^0 &= \{(z,t) \in U : |\tau| < |\xi|\}. \end{aligned}$$

Then the principal symbol of $\bar{\partial}_b \bar{\partial}_b^*$ does not vanish in Γ^0, so that $\bar{\partial}_b \bar{\partial}_b^*$ is microlocally AH there. In Γ^-, $\bar{\partial}_b$ is subelliptic, and is of principal type, hence is microlocally AH. Hence if $\bar{\partial}_b \bar{\partial}_b^* u = 0$, then $\bar{\partial}_b^* u$ is analytic in $\Gamma^0 \cup \Gamma^-$. Moreover, $\bar{\partial}_b^*$ is microlocally AH in Γ^+, being subelliptic and of principal type there. Thus $\bar{\partial}_b$ is RAH at 0 if and only if $\bar{\partial}_b \bar{\partial}_b^*$ is microlocally AH in Γ^+; relative analytic hypoellipticity reduces to a more standard notion.

Our best guess concerning analytic hypoellipticity of $\bar{\partial}_b$ for this class of examples, assuming always that P is subharmonic but not harmonic:

Conjecture 2.2. $\bar{\partial}_b$ *is relatively analytic hypoelliptic at* $\zeta = (z,t)$ *if and only if either* $\triangle P(z) \neq 0$, *or* z *is an isolated zero of* $\triangle P$.

When $\triangle P(z) \neq 0$, relative analytic hypoellipticity was established by Geller [G]; it follows alternatively from [Tv2], using the remark above on Γ^+. That analytic hypoellipticity should break down when the zeroes are not isolated, is a special case of a belief expressed in [Tv2]. The related operator

$$\mathcal{L} = -X^2 - Y^2$$

has been more intensively studied, and the same conjecture can be advanced for it. Only fragmentary results seem to be known in the weakly pseudoconvex, or non-symplectic, case:

(1) $\mathcal{L}$ is AH when $P(x,y) = x^{k+2} + y^{\ell+2}$, when k, ℓ are nonnegative even integers [GS][1].

(2) $\mathcal{L}$ fails to be AH when $P(x,y) = x^m$, $m = 3,4,5,\ldots$[2]. See [He],[PR] for $m = 3$, [HH] for $m = 5,7,\ldots$ and most recently [C3] for $m = 4,6,8,\ldots$.

(3) $\bar{\partial}_b$ fails to be RAH when $P(x,y) = Q(x)$, wherever the second derivative Q'' vanishes. The special case $Q(x) = x^m$ was treated in [CG] and is the starting point for the present article; the general case is implicit in [C4].

(4) $\bar{\partial}_b$ is RAH when P is a homogeneous polynomial in x, y (with respect to the isotropic dilation group $(x,y) \mapsto (rx, ry)$), whose Laplacian vanishes only at the origin, assuming always that P is subharmonic. A proof is sketched in the second paragraph of §7.

The formulation of the conjecture for $\bar{\partial}_b$ on a general three-dimensional CR manifold is:

Conjecture 2.3. *Let M be an analytic three-dimensional CR manifold, pseudoconvex and of finite type. Then $\bar{\partial}_b$ fails to be relatively analytic hypoelliptic at $\zeta \in M$ if and only if, for every neighborhood U of ζ, there exists $\varepsilon > 0$ and a nonconstant smooth curve $\gamma : (-\varepsilon, \varepsilon) \mapsto U$, such that for every t,*

$$\gamma(t) \text{ is a weakly pseudoconvex point of } M$$

and

$$\gamma'(t) \in T^{1,0}M \oplus T^{0,1}M.$$

The breakdown of relative analytic hypoellipticity has been established for Reinhardt domains in $\mathbb{C}^2$ satisfying this condition [C4].

---

[1] In [GS] the following condition is proved to be sufficient: $\triangle P$ vanishes only at the origin, and $\triangle P(r^{1/k}x, r^{1/\ell}y) \equiv r \triangle P(x,y)$.

[2] Here the standing hypothesis of subharmonicity is momentarily dropped.

This geometric condition may be reformulated in terms of the properties of contact of M with analytic discs. If M is realized locally as an embedded real hypersurface in $\mathbb{C}^2$, and if a curve γ of the type described exists, then there exists an analytic disc $\mathcal{D} \subset \mathbb{C}^2$ which intersects M in a subarc of γ. Conversely whenever an analytic disc intersects M in a real curve, that curve is of the type described. One may go further to conjecture that the range of s for which relative G^s hypoellipticity holds, is related to the order of tangency to M of such a disc, in the direction transverse to γ at $\mathcal{D} \cap M$. The higher the order of contact in the transverse direction, the larger should be the critical value of s below which G^s hypoellipticity fails.

It should be emphasized that the notion of analytic hypoellipticity in question is a local one. There is evidence that global regularity is a different story altogether.

Conjecture 2.4. *Let M be a compact, analytic three-dimensional CR manifold (without boundary) of finite type. If $u \in L^2(M)$ and $\bar{\partial}_b \bar{\partial}_b^* u \in C^\omega(M)$, then $\bar{\partial}_b^* u \in C^\omega(M)$.*

An analogous conjecture, concerning the $\bar{\partial}$–Neumann problem on a pseudoconvex, bounded domain $\Omega \subset \mathbb{C}^2$ with analytic boundary, is equally plausible. For circular domains, the conjecture has been proved [Ch]. In particular, one has global analytic hypoellipticity for the $\bar{\partial}$–Neumann problem, for those same Reinhardt domains for which $\bar{\partial}_b$ has been proved not to be RAH. A related question is whether the Bergman kernel $B(z, w)$ is analytic on $\Omega \times \bar{\Omega}$.

3. Partial reduction to $\mathbb{C}^1$.

Letting τ be a variable dual to t, and taking a partial Fourier transform in t, reduces $\bar{\partial}_b$ to a one-parameter family of operators, acting on functions of $z \in \mathbb{C}$:

$$D_\tau = \bar{\partial} + \tau \cdot (\bar{\partial} P)$$

where $(\bar{\partial} P)$ operates by multiplication. Then $\bar{\partial}_b^*$ goes over to D_τ^*, and we are interested in $D_\tau \circ D_\tau^*$. Because the only issue is the region Γ^+, we restrict attention to $\tau > 0$.

Analytic hypoellipticity of $\bar{\partial}_b$ is related to the following property of $\{D_\tau\}$:

Definition 3.1. *We say that solutions of $D_\tau D_\tau^*$ are exponentially small at z if for any neighborhood $U \ni z$ there exist $C < \infty$, $\varepsilon > 0$ such that whenever $\tau \geq 1$, $u \in L^2(U)$ and $D_\tau D_\tau^* u \equiv 0$ in U, one has*

$$|u(z)| \leq C e^{-\varepsilon \tau} \|u\|_{L^2(U)}.$$

If $\bar{\partial}_b$ is RAH near (z, t), then solutions of $D_\tau D_\tau^*$ must be exponentially small at z. For let τ and a solution u of $D_\tau D_\tau^*$ be given. Define $f(z, t) = u(z) e^{it\tau}$. Then $\bar{\partial}_b \bar{\partial}_b^* f \equiv 0$. Assuming $\bar{\partial}_b$ to be RAH in $U \times (-1, 1)$, it would follow that for all k,

$$|\frac{\partial^k}{\partial t^k} \bar{\partial}_b^* f(z, t)| \leq C^{k+1} k! \|f\|_{L^2(U \times (-1, 1))}.$$

Normalize so that $\|u\|_{L^2(U)} = 1$. Then since $\bar{\partial}_b^* f(z,t) = e^{it\tau} D_\tau^* u(z)$,

$$|D_\tau^* \dot{u}(z)| \leq \min_k \left(C^{k+1} k! \tau^{-k} \right) \leq Ce^{-\varepsilon\tau}$$

for some $\varepsilon > 0$. The same bound holds in a neighborhood of z.

It remains to deduce that u itself satisfies an exponential decay estimate; this is related to the analytic hypoellipticity of $\bar{\partial}_b^*$ in Γ^+.

Lemma 3.2. *Suppose that $\tau \geq 1$, that $\|u\|_{L^2(U)} = 1$, that P is subharmonic and that $D_\tau D_\tau^* u \equiv 0$ in $U \ni z$. Suppose further that*

$$\sup_{\zeta \in U} |D_\tau^* u(\zeta)| \leq Ce^{-\varepsilon\tau}.$$

Then

$$|u(z)| \leq C' e^{-\varepsilon'\tau}$$

where $C' < \infty$, $\varepsilon' > 0$ depend only on C, ε, P.

In other words, solutions of D_τ^* are always exponentially small, without hypothesis on the zeroes of ΔP. We include an elementary proof, partly in order to indicate one difference between operators of principal type, such as $\bar{\partial}_b^*$, and operators with multiple characteristics such as $\bar{\partial}_b \bar{\partial}_b^*$.

Proof. Write $D_\tau^* u = -e^{\tau P} \partial e^{-\tau P} u$, where $\partial = \partial/\partial z$. Note that if h is any harmonic function, if h' denotes a harmonic conjugate, and if $\tilde{P}$ is defined to be $P - h$, then

$$D_\tau^* = -e^{i\tau h'} \left[e^{\tau \tilde{P}} \partial e^{-\tau \tilde{P}} \right] e^{-\tau i h'},$$

since $\partial(e^{\tau[h - ih']}) = 0$. Thus if h, h' are taken to be real-valued, then P may be replaced by $\tilde{P}$ in the definition of D_τ^*, at the minor expense of conjugation with a unitary operator.

Suppose, to begin, that $\Delta P(z) \neq 0$. Choose a real, harmonic, quadratic polynomial $h(x,y)$ so that $\tilde{P} = P - h$ satisfies $\tilde{P}(z) = 0$, $\nabla\tilde{P}(z) = 0$, and so that the Hessian of $\tilde{P}$ at z is positive definite. Let h' be a real-valued harmonic conjugate, and set $\tilde{u} = e^{-i\tau h'} u$ and $D = \left[e^{\tau \tilde{P}} \partial e^{-\tau \tilde{P}} \right]$, so that $D\tilde{u}(\zeta) = O(e^{-\varepsilon\tau})$ for $|z - \zeta| < 2\delta$. If δ is chosen to be sufficiently small, then $\tilde{P}(\zeta) > c\delta^2$ for $\delta < |z - \zeta| < 2\delta$, so that (in L^2 norm)

$$e^{-\tau \tilde{P}} \tilde{u}(\zeta) = O(e^{-c\tau\delta^2}) \qquad \text{when } \delta < |z - \zeta| < 2\delta.$$

We still have

$$\partial(e^{-\tau \tilde{P}} \tilde{u}) = O(e^{-\varepsilon\tau}) \qquad \text{when } |z - \zeta| < 2\delta.$$

Stokes' theorem gives $\tilde{u}(\zeta) = O(e^{-\tau\varepsilon'})$ for $|z - \zeta| < \delta/4$, where $\varepsilon' = \min(\varepsilon, c\delta^2)$. But $\tilde{u}(z) = u(z)$.

Next suppose that ΔP has an isolated zero at z. By the case just treated, for all sufficiently small $\delta > 0$ one has $u(\zeta) = O(e^{-\varepsilon(\delta)\tau})$ when $\delta < |z - \zeta| < 2\delta$. Fix such a δ. We require an elementary estimate for D_τ^*:

Lemma 3.3. *For any compact subset $K \subset \mathbb{C}$ of the domain of definition of P, there exists $C < \infty$ such that for any $g \in C_0^1$ supported in K, for all $\tau \geq 1$,*

$$\|g\|_{L^2} \leq C\|D_\tau^* g\|_{L^2}.$$

In fact, $C \leq C' \tau^{-\varepsilon}$ for some $\varepsilon > 0$; we accept the lemma and defer to [C1] or [C2] for proof.

Let $\eta \in C_0^\infty$ be identically one for $|z - \zeta| \leq \delta$ and be supported where $|z - \zeta| < 2\delta$. Apply the lemma to obtain

$$\begin{aligned}
\|\eta u\|_{L^2} &\leq C\|D_\tau^*(\eta u)\|_{L^2} \\
&\leq C\|\eta D_\tau^* u\|_{L^2} + C\|u \nabla \eta\|_{L^2} \\
&\leq C e^{-\varepsilon'\tau}.
\end{aligned}$$

Since D_τ, D_τ^* are elliptic operators whose coefficients are $O(\tau)$, and since $D_\tau D_\tau^* u \equiv 0$, it follows that for some finite N, $u(z) = O(\tau^{-N} e^{-\varepsilon'\tau}) = O(e^{-\varepsilon''\tau})$. Note that this argument applies equally well, given instead that z is an isolated point of the set on which u is not known to satisfy the exponential decay estimate.

Next consider the case where $\Delta P(z) = 0$ and where the intersection of $\{\Delta P = 0\}$ with some neighborhood of z is a smooth curve, γ. By conjugating with a unitary operator as above, we may assume that $P(z) = 0 = \nabla P(z) = \nabla^2 P(z)$. Fixing a sufficiently small $\delta > 0$, one may choose a homogeneous quadratic, harmonic real-valued polynomial h such that in a neighborhood of $\gamma \cap \{\delta \leq |z - \zeta| \leq 2\delta\}$, $h(\zeta)$ is strictly positive. Outside that neighborhood, $u(\zeta) = O(e^{-\varepsilon\tau})$ where $\varepsilon > 0$. If $r > 0$ is chosen to be so small that the exponential decay estimate is not destroyed outside the neighborhood in question, then $e^{-r\tau h} u = O(e^{-\varepsilon'\tau})$ in the annulus $\delta < |z - \zeta| < 2\delta$, for some $\varepsilon' > 0$. The argument of the last paragraph now applies.

The only remaining case is where $\Delta P(z) = 0$, z is not an isolated point of ΔP, and the zero variety of ΔP in a neighborhood of z does not form a smooth curve. The set of all such z is discrete, so that the exponential decay estimate holds at every point of some neighborhood of z, save possibly z itself. Therefore it holds at z, as remarked two paragraphs ago. ∎

Analogous to our surmise regarding relative analytic hypoellipticity of $\bar{\partial}_b$ is the

Conjecture 3.4. *Solutions of $D_\tau D_\tau^*$ are exponentially small, if and only if all zeroes of ΔP are isolated.*

One half of the conjecture is straightforward.

Proposition 3.5. *If all zeroes of ΔP are isolated, then solutions of $D_\tau D_\tau^*$ are exponentially small.*

Proof. We take for granted the known fact that $\bar{\partial}_b$ is relatively analytic hypoelliptic at strictly pseudoconvex points, that is, at (z,t) where $\Delta P(z) \neq 0$. Hence solutions of $D_\tau D_\tau^*$ are exponentially small near all such z.

Suppose that $D_\tau D_\tau^* u = 0$. If u is exponentially small near ζ, it follows that $D_\tau u$ and $D_\tau^* u$ are also exponentially small near ζ, using ellipticity plus the fact that the coefficients of these operators are only $O(\tau)$.

Suppose that z is an isolated zero of ΔP, and choose $\delta > 0$ so that $\Delta P(\zeta) > 0$ whenever $0 < |z - \zeta| \leq 2\delta$. The analogue of Lemma 3.3 holds with D_τ^* replaced by $D_\tau D_\tau^*$ [C1], [C2]. Let $\eta \in C_0^\infty$ be identically one for $|z - \zeta| \leq \delta$ and be supported in the concentric ball of radius 2δ. Then $D_\tau D_\tau^*(\eta u)$ is supported where $\delta \leq |z - \zeta| < 2\delta$, and is exponentially small. Therefore $\|\eta u\|_{L^2}$ is exponentially small, whence the same follows for $u(z)$.

4. The $\bar{\partial}$ equation in weighted L^2 spaces.

The considerations of the preceding section motivate a more detailed study of D_τ and $\mathcal{D} = D_\tau D_\tau^*$. Since $D_\tau = e^{-\tau P} \circ \bar{\partial} \circ e^{\tau P}$, studying D_τ in L^2 is equivalent to studying $\bar{\partial}$ in $L^2_{\tau P} = \{f : \int_{\mathbb{C}} |f|^2 e^{-2\tau P} < \infty\}$. However, certain results may be more simply formulated for D_τ than for $\bar{\partial}$, so we continue to work with the former. We shall first present a somewhat technical result, then explain its connection with the question of exponential smallness of solutions of $\mathcal{D}$.

The following seems to be a reasonable degree of generality in which to study D_τ. Let $\varphi : \mathbb{C} \mapsto \mathbb{R}$ be a subharmonic, nonharmonic function. Its Laplacian is then a locally finite measure, which we denote by μ. Assume that μ is a doubling measure, that is, there exists $C < \infty$ such that

$$(4.1) \qquad \mu(B(z, 2r)) \leq C\mu(B(z, r)) \qquad \forall z \in \mathbb{C}, r > 0.$$

Assume further that there exists $\delta > 0$ such that

$$(4.2) \qquad \mu(B(z, 1)) \geq \delta \qquad \forall z \in \mathbb{C}.$$

Any subharmonic, nonharmonic polynomial P is an example of such a φ; indeed this is the case of primary interest for us.

Define $D = \bar{\partial} + \varphi_{\bar{z}}$ where $\varphi_{\bar{z}} = \bar{\partial}\varphi$, $D^* = -\partial + \varphi_z$, and $\mathcal{D} = DD^*$. Write L^2 to denote $L^2(\mathbb{C})$ with respect to Lebesgue measure, and $\|\cdot\|$ to denote the corresponding norm.

A preliminary result is

Lemma 4.1. *Suppose that φ satisfies hypotheses (4.1) and (4.2). Then*

(1) *D^*, D and $\mathcal{D}$ may be regarded as densely defined, closed linear operators on L^2.*

(2) There exists $C < \infty$ such

$$\|u\| \leq C\|D^*u\| \text{ for all } u \text{ in the domain of } D^*.$$

(3) The domain of $\mathcal{D}$ contains the domains of D and D^*, and there exists $C < \infty$ such that for all u in the domain of $\mathcal{D}$,

$$\|u\| + \|Du\| + \|D^*u\| \leq C\|\mathcal{D}u\|.$$

(4) For every $f \in L^2$ there exists a unique u in the domain of $\mathcal{D}$ such that $\mathcal{D}u = f$. $g = D^*u$ is the unique solution of $Dg = f$ which is orthogonal to the nullspace of D.

(5) The nullspace of D in $L^2(\mathbb{C})$ has infinite dimension.

Let G denote the distribution-kernel for the bounded, linear operator $\mathcal{D}^{-1}$. Let S denote the distribution-kernel for the orthogonal projection of L^2 onto the nullspace of D. Fix a function $\varrho : \mathbb{C} \mapsto \mathbb{R}^+$ satisfying

$$C_0^{-1} < \mu(B(z, \varrho(z))) < C_0 \qquad \forall z \in \mathbb{C}$$

for some large constant C_0; ϱ exists and may be chosen to be C^∞, by (4.1) and (4.2). Define a Riemannian metric by

$$d\rho^2 = \varrho^{-2} ds^2,$$

where ds^2 is the Euclidean metric. ϱ^{-2} is a smoothed-out version of ΔP; it is comparable to ΔP where ΔP is large, but is strictly positive where $\Delta P = 0$. Because of the hypotheses on μ, there exists $\delta > 0$ such that $\rho(z, \zeta) \geq \delta|z - \zeta|$ for all $z, \zeta \in \mathbb{C}$. Roughly speaking, ρ will be much larger than the Euclidean distance, where $\Delta\varphi$ is large.

Our principal result regarding $\mathcal{D}$ and D is:

Theorem 4.2. *Suppose that φ satisfies (4.1) and (4.2). Then G, S are continuous functions away from the diagonal. There exist $\varepsilon > 0$, $C < \infty$ such that*

(4.3)
$$|G(z, \zeta)| \leq C \begin{cases} \log(2\varrho(z)/|z - \zeta|) & \text{for } |z - \zeta| \leq \varrho(z) \\ \exp(-\varepsilon\rho(z, \zeta)) & \text{for } |z - \zeta| \geq \varrho(z) \end{cases}$$

and

(4.4)
$$|S(z, \zeta)| \leq C\varrho(z)^{-2} \exp(-\varepsilon\rho(z, \zeta)) \qquad \forall z, \zeta \in \mathbb{C}.$$

The constants C, ε depend only on the constant in (4.1).

Corollary 4.3. *Let* φ *satisfy* (4.1) *and* (4.2). *Let* $p \in [1, \infty]$ *and suppose that* $fe^{-\varphi} \in L^p(\mathbb{C})$. *Then there exists a solution of* $\bar{\partial} u = f$ *on* $\mathbb{C}$, *satisfying* $ue^{-\varphi} \in L^p(\mathbb{C})$.

This is proved by solving $\mathcal{D}g = fe^{-\varphi}$. (4.3) and a similar estimate for the distribution-kernel of $D^* \circ \mathcal{D}$ imply that $g, D^*g \in L^p$, and $u = e^\varphi D^*g$ is then a solution with the desired size restriction.

The estimate for G is sharp near the diagonal, more precisely when $|z - \zeta| \leq \varepsilon_0 \varrho(z)$. Since $\rho(z, \zeta) \geq \delta|z - \zeta|$, both G and S decay at least like $\exp(-\varepsilon|z - \zeta|)$ as $|z - \zeta| \to \infty$. However, $\rho(z, \zeta)$ could grow like some power of $|z - \zeta|$, constrained only by the constant in (4.1). How much *faster* than exponentially do G and S decay? This question is subtler, is intimately linked with that of exponential smallness of solutions of the one-parameter family of operators obtained by replacing φ by $\tau\varphi$, and is thereby linked with analytic hypoellipticity.

To see the connection, consider $\varphi = \tau P$, where P is a subharmonic, nonharmonic polynomial. Suppose moreover that P is homogeneous of some degree $m \in \{2, 4, 6, \dots\}$. Consider $D_\tau, D_\tau^*, \mathcal{D}_\tau = D_\tau D_\tau^*$ as before; there are distribution-kernels G_τ and S_τ. These are related by:

$$(4.5) \qquad G_\tau(z, \zeta) = G_1(\tau^{1/m}z, \tau^{1/m}\zeta), \qquad S_\tau(z, \zeta) = \tau^{2/m} S_1(\tau^{1/m}z, \tau^{1/m}\zeta).$$

$\mathcal{D}_\tau G_\tau \equiv 0$ off of the diagonal, with $\mathcal{D}_\tau$ acting in the z variable. Moreover, the Szegö projection is $I - D_\tau^* \mathcal{D}_\tau^{-1} D_\tau$ where I denotes the identity. Therefore if solutions of $\mathcal{D}_\tau$ are exponentially small, $|S_\tau(z, 0)| + |G_\tau(z, 0)| = O(e^{-\varepsilon\tau})$ for $|z| \sim 1$. By (4.5), this means that

$$|S_1(z, 0)| + |G_1(z, 0)| = O(\exp(-\varepsilon|z|^m)) \qquad \text{as } |z| \to \infty.$$

How well does Theorem 4.2 accord with experiment? Let us focus on the Szegö kernel S_1. When $P(z) = |z|^m$, $m = 2, 4, 6, \dots$, it yields

$$S_1(z, 0) = O(\exp(-\varepsilon|z|^{m/2})).$$

But solutions of $\mathcal{D}$ are exponentially small in these cases (for instance, by Proposition 3.5), so the correct estimate is $\exp(-\varepsilon|z|^m)$; we are off by a factor of two in the exponent, for all m.

The proof of Theorem 4.2 *is* actually sharp, for a different class of operators, namely Schrödinger operators $L_\tau = -\triangle + \tau\triangle P$. These may be analyzed in the same way as $\mathcal{D}$, and have fundamental solutions which satisfy the same estimates as G. Analytic hypoellipticity fails for operators $-\triangle_{x,y} - \triangle P(x, y)\partial^2/\partial t^2$; solutions of L_τ need not be exponentially small as $\tau \to +\infty$; and the exponent $m/2$ in the decay estimate for the fundamental solution is sharp. This last may be seen on an intuitive level by applying the Feynmann-Kac formula to e^{-sL_1}; for $P(x,y) = x^m + y^m$, it may be checked by examining the associated heat kernel by separation of variables.

Thus it might appear that, speaking roughly, the estimates (4.3) and (4.4) are always off by a factor of two in the exponent. But consider $P(x,y) = x^m$, $m = 2, 4, \ldots$. Then (4.4) yields $S_1(x + i0, 0) = O(\exp(-\varepsilon|x|^{m/2}))$, but only $S_1(0 + iy, 0) = O(\exp(-\varepsilon|y|^1))$. Either the exponent is off by a factor of m along the y axis, or solutions of $D_\tau D_\tau^*$ fail to be exponentially small. In fact, we will shortly see that for $m = 4, 6, \ldots$, $S_1(0 + iy, 0)$ is not $O(\exp(-\varepsilon|y|^q))$ for any $q > 1$; (4.4) gives the optimal rate of exponential decay in this direction, for even $m \geq 4$ (though not for $m = 2!$).

5. Non-analytic Szegö kernels.

We have seen reason to suspect that $\bar{\partial}_b$ should fail to be RAH for the CR manifolds $M_m = \{z \in \mathbb{C}^2 : \Im(z_2) = [\Re(z_1)]^m\}$, for $m = 4, 6, 8, \ldots$. This is reflected in the failure of the associated Szegö kernel to be analytic at certain points off of the diagonal. Recall the identification with $\mathbb{C} \times \mathbb{R}$ introduced earlier. The Szegö projection is the orthogonal projection of $L^2(\mathbb{C} \times \mathbb{R})$ onto the intersection of L^2 with the nullspace of $\bar{\partial}_b = \bar{\partial} - i(\bar{\partial}P)\partial_t$.

The Szegö projection has a distribution-kernel $S((z,t);(w,s))$, which is known to be a C^∞ function off of the diagonal. For M_m it has been computed explicitly [N]; for our purpose it will suffice to specialize to $(w,s) = 0$. Then

$$(5.1) \qquad S((z,t);0) = (2\pi)^{-2} \int_0^\infty e^{it\tau} \tau^{2/m} e^{-\tau x^m} \int_{-\infty}^\infty e^{\tau^{1/m}\eta z} N(\eta)^{-1} \, d\eta \, d\tau$$

where $z = x + iy$ and

$$(5.2) \qquad N(\zeta) = \int_{-\infty}^\infty e^{2(\zeta s - s^m)} \, ds.$$

Equivalently, the Szegö kernel S_τ for the orthogonal projection of $L^2(\mathbb{C})$ onto the nullspace of D_τ satisfies

$$(5.3) \qquad S_\tau(z,0) = (2\pi)^{-1} \tau^{2/m} e^{-\tau x^m} \int_{-\infty}^\infty e^{\eta\tau^{1/m}z} F(\eta)^{-1} \, d\eta.$$

If $S_1(0 + iy, 0) = O(\exp(-\varepsilon|y|^q))$ for some $q > 1$, then $\int_{-\infty}^\infty \exp(-i\eta y) S_1(0 + iy, 0) \, dy$ would extend to an entire function of $\eta \in \mathbb{C}$. By (5.3), this last expression is simply $N(\eta)^{-1}$ for $\eta \in \mathbb{R}$. Thus N must extend to a *nonvanishing* entire function.

Analyzing the integral defining N by the method of (real) stationary phase reveals that N is an entire function of order precisely $m/(m - 1)$. In particular, $|N(\zeta)| \geq \varepsilon \exp(\varepsilon|\zeta|^{m/(m-1)})$ for real ζ. When $m = 2$, N is a Gaussian function, and has no zeroes. We already know that this must be so; in this case $\Delta P(z) \equiv 2$ does not vanish, so

$\bar{\partial}_b$ is RAH. $S(\cdot, 0)$ belongs to the range of $\bar{\partial}_b^*$ and is annihilated by $\bar{\partial}_b$, away from 0; when $\bar{\partial}_b$ is RAH, it must therefore be analytic except at 0. But when $m = 4, 6, \ldots, m/(m-1)$ is no longer an integer, and any entire function of non-integral order must have infinitely many zeroes. Thus we have:

Proposition 5.1. [CG] *The Szegö kernel for the CR manifolds $\{\Im(z_2) = [\Re(z_1)]^m\}$, for $m = 4, 6, 8, \ldots$, fails to be analytic at certain points off of the diagonal. $\bar{\partial}_b$ is not relatively analytic hypoelliptic, and solutions of $D_\tau D_\tau^*$ need not be exponentially small.*

Moreover, the exponential decay estimate (4.4) for S_1 is sharp along the y axis. It is sharp in other cases, as well. Consider $\varphi(x, y) = b(x)$ where $1 \leq b''(x) \leq 2$ for all x, and let Σ denote temporarily the distribution-kernel for the orthogonal projection onto the nullspace of D. Then (4.4) yields $\Sigma(0 + iy, 0) = O(\exp(-\varepsilon|y|^1))$, and the exponent 1 is optimal [C2] except when b'' is constant, in which case the exponent should be 2.

In the scale of Gevrey classes, S belongs to G^s for all $s \geq m$, but no better [CG]. We shall digress in the next section to understand the breakdown of relative analytic hypoellipticity for $\bar{\partial}_b$ in a more direct fashion, then return to the Szegö kernel in §7.

6. Singular solutions.

Continue to assume that $P(x, y) = x^m$. Applying the partial Fourier transform in the y and t variables to $\bar{\partial}_b \bar{\partial}_b^*$ yields the two-parameter family of operators

$$[-\frac{d}{dx} + (\eta - \tau m x^{m-1})] \circ [\frac{d}{dx} + (\eta - \tau m x^{m-1})].$$

As above, the general case $\tau > 0$ reduces via a dilation of the x variable to $\tau = 1$. Thus set

$$L_\zeta = [-\frac{d}{dx} + (\zeta - m x^{m-1})] \circ [\frac{d}{dx} + (\zeta - m x^{m-1})]$$

where ζ is a complex parameter. Since L_ζ factors as a product of two first-order operators, one finds an explicit solution of $L_\zeta f_\zeta = 0$:

(6.1)
$$f_\zeta(x) = e^{-\zeta x + x^m} \int_{-\infty}^{x} e^{2(\zeta s - s^m)} \, ds.$$

Fix ζ. Then formally,

(6.2)
$$F(z, t) = \int_0^\infty e^{it\tau} e^{i\tau^{1/m}\zeta y} f_\zeta(\tau^{1/m} x) \, d\tau$$

satisfies $\bar{\partial}_b \bar{\partial}_b^* F \equiv 0$, and

$$\bar{\partial}_b^* F(z, t) = \int_0^\infty e^{it\tau} e^{-\tau x^m} e^{z\tau^{1/m}\zeta} \, d\tau.$$

If ζ has positive imaginary part σ and if f_ζ is a bounded function of x on the whole real line[3], then these integrals converge absolutely in $\{y > 0\}$. Moreover

$$\frac{\partial^k}{\partial t^k}\bar\partial_b^* F(0+i,0) = i^k \int_0^\infty \tau^k e^{-\tau^{1/m}\sigma}\, d\tau$$

$$= i^k m\sigma^{-mk-m}\Gamma(mk+m).$$

Hence $\bar\partial_b^* F$ would be not be analytic, and we would have a counterexample. If ζ had negative imaginary part, the same reasoning would apply for $y < 0$. Note that when $\zeta \in \mathbb{R}$, $f_\zeta(x)$ behaves like a positive constant times $\exp(-\zeta x + x^m)$ as $x \to +\infty$; it certainly does not remain bounded.

Consider a simpler problem: for the example

(6.3)
$$\frac{\partial^2}{\partial x_1^2} + x_1^2 \frac{\partial^2}{\partial x_2^2} + \frac{\partial^2}{\partial x_3^2}$$

in $\mathbb{R}^3$, the same procedure leads to the problem of finding $\zeta \in \mathbb{C}$ and a bounded solution f of

$$\left(-\frac{d^2}{dx^2} + x^2 + \zeta^2\right)f = 0.$$

This may be done explicitly [H, p. 310]: taking λ^2 to be an eigenvalue of $-d^2/dx^2 + x^2$, f to be a corresponding eigenfunction and $\zeta = i\lambda$, yields a solution. Proceeding as in the last paragraph produces a non-analytic solution to (6.3).

Returning to the L_ζ, we do not know explicit values of ζ for which f_ζ remains bounded. Clearly f_ζ is C^∞, and tends to zero as $x \to -\infty$, for any ζ. For f_ζ to remain bounded as $x \to +\infty$, a necessary and sufficient condition is that

$$\int_{-\infty}^\infty e^{2(\zeta s - s^m)}\, ds = 0;$$

this is precisely the condition $N(\zeta) = 0$ which arose in the analysis of the Szegö kernel! Therefore there do exist values of ζ for which f_ζ remains bounded, and (6.2) constructs comparatively simple solutions F to $\bar\partial_b\bar\partial_b^*$ with $\bar\partial_b^* F$ non-analytic.

This procedure is not completely satisfactory. Consider $\mathcal{L} = -X^2 - Y^2$, where $\bar\partial_b = X + iY$. Then $\mathcal{L}$ and $\bar\partial_b\bar\partial_b^*$ are closely related. For instance, denoting by $X_\tau, Y_\tau, \mathcal{L}_\tau$, the images under the partial Fourier transfom in the t variable, the associated quadratic forms satisfy

$$\langle \mathcal{L}_\tau f, f\rangle \le \langle D_\tau D_\tau^* f, f\rangle \le 2\langle \mathcal{L}_\tau f, f\rangle$$

---

[3] It suffices to have sub-exponential growth, but it happens that f_ζ either grows essentially like $\exp(x^m)$, or decays rapidly as $x \to +\infty$.

for all $\tau > 0$ and $f \in C_0^2(\mathbb{C})$, as follows from an integration by parts. Yet the procedure just indicated for $\bar{\partial}_b \bar{\partial}_b^*$ does not apply immediately to $\mathcal{L}$; one seeks bounded solutions f for

$$\Lambda_\zeta = -\frac{d^2}{dx^2} + (\zeta - mx^{m-1})^2,$$

but is no longer able to exhibit such solutions in closed form.

Nonetheless, solutions to Λ_ζ which behave like the f_ζ do exist. Set $\gamma = -(m-1)/2$ and $\Phi_\zeta(x) = \zeta x - x^m$.

Lemma 6.1. [C3] *Let $m \in \{4, 6, 8, \dots\}$. For each $\zeta \in \mathbb{C}$ there exist unique solutions $f_\zeta^\pm$ of $\Lambda_\zeta f_\zeta^\pm = 0$ satisfying*

$$f_\zeta^\pm(x) = e^{\Phi_\zeta(x)}|x|^\gamma + O(|e^{\Phi_\zeta(x)}| \cdot |x|^{\gamma-1})$$

as $x \to \pm\infty$, respectively. They depend holomorphically on ζ. Their Wronskian $W(x, \zeta)$ depends only on ζ, and is an entire function of order exactly $m/(m-1)$.

If $W(\zeta) = 0$, then the $f_\zeta^\pm$ are linearly dependent. Thus f_ζ^+ remains bounded both as $x \to +\infty$ and as $x \to -\infty$. The same construction as before produces non-analytic solutions for $\mathcal{L}$. Thence

Proposition 6.2. $\mathcal{L} = -\partial_x^2 - (\partial_y - mx^{m-1}\partial_t)^2$ *is not analytic hypoelliptic, for $m = 4, 6, \dots$.*

A variant of the lemma works for $m = 3, 5, 7, \dots$ as well, so the corresponding operators $\mathcal{L}$ again fail to be analytic hypoelliptic, as was first proved in [He],[PR],[HH].

7. More on the Szegö kernel.

To a limited extent, the non-analytic Szegö kernel for M_m may be understood in terms of substantially simpler non-analytic functions; S admits an expansion in terms of functions of the type constructed in §6. Observe first that $S(z, t) = S((z, t), 0)$ is analytic except where $t = 0 = x$. For in addition to being annihilated by $\bar{\partial}_b$, S satisfies an additional first-order equation, which expresses the fact that $S(rz, r^m t) \equiv r^{-2-m} S(z, t)$. These two constitute an elliptic first-order system (that is, their principal symbols have no common zeroes), except where $t = 0$, so S is analytic except possibly where $t = 0$. But taking the partial Fourier transform in t, we know already that solutions of $D_\tau D_\tau^*$ are exponentially small where $x \neq 0$. Reversing the Fourier transform, it follows that

$$\left|\frac{\partial^k}{\partial t^k} S(z, t)\right| \leq C(z, t)^{k+1} k! \qquad \forall k$$

for $x \neq 0$. Combining this with the fact that S is microlocally analytic outside Γ^+, and with the FBI transform, it follows that S is analytic where $x \neq 0$. Therefore the singularity of S is found only on the line $x = t = 0$.

This type of argument also applies to the relative fundamental solution to $\bar{\partial}$, whenever $P(x, y)$ is a homogeneous polynomial. If in addition ΔP vanishes only at the origin, it follows that the relative fundamental solution (as a function on $[\mathbb{C} \times \mathbb{R}] \times [\mathbb{C} \times \mathbb{R}]$) is analytic off the diagonal, hence that $\bar{\partial}_b$ is RAH.

Since $S(x - iy, -t) = \overline{S(x + iy, t)}$, and since S is homogeneous, the singularity actually occurs along the entire line in question. We seek to describe that singularity, and in doing so, sacrifice no generality in assuming that $y > 0$ and that $|x|$, t^m are small in comparison with y.

Recall the entire function N defined in (5.2). Let $m \in \{4, 6, 8, \dots\}$ and let $\{\zeta_\nu = \lambda_\nu + \sigma_\nu\}$ be the collection of all its zeroes. Set $\gamma = (m - 2)/2(m - 1)$.

Lemma 7.1. N has only finitely many zeroes in any strip $\{\zeta \in \mathbb{C} : |\Im(\zeta)| \leq A < \infty\}$. Moreover there exists $\varepsilon > 0$ such that

$$|\sigma_\nu| \geq \varepsilon |\lambda_\nu|^\gamma \qquad \forall \nu.$$

This is a consequence of the method of stationary phase.

Define $Q_\nu(z)$ to be the residue of the function $\zeta \mapsto \exp(iz(\zeta - \zeta_\nu))N(\zeta)^{-1}$ at $\zeta = \zeta_\nu$; Q_ν is a polynomial in z. Define

$$S_\nu(z, t) = (2\pi)^{-2} \int_0^\infty e^{it\tau} e^{-\tau x^m} Q_\nu(\tau^{1/m} z) e^{\tau^{1/m} \zeta_\nu z} \tau^{2/m} \, d\tau.$$

When $\sigma_\nu > 0$, the integral converges absolutely for $y > 0$ and defines a C^∞ function. These are slight generalizations of the singular solutions $\bar{\partial}_b^* F$ constructed in §6; the additional factors of Q_ν are the only significant alteration.

S is built up, in a certain sense, of the S_ν. Suppose that $\sigma > 0$ and that none of the σ_ν equal σ. Define

$$(7.1) \qquad E_\sigma(z, t) = (2\pi)^{-2} \int_0^\infty e^{it\tau} e^{-\tau x^m} \tau^{2/m} \int_{-\infty}^\infty e^{\tau^{1/m}(\eta + i\sigma)z} F(\eta + i\sigma)^{-1} \, d\eta \, d\tau.$$

The integral may still be shown to be converge absolutely where $y > 0$ and to define a C^∞ function there.

Proposition 7.2. Suppose that σ does not equal the imaginary part of any zero of N. Then where $y > 0$,

$$S((z, t), 0) = \sum_{0 < \sigma_\nu < \sigma} S_\nu(z, t) + E_\sigma(z, t).$$

This is obtained merely by shifting the contour of integration in (5.1) from $\mathbb{R}$ to $\mathbb{R} + i\sigma$, bearing in mind that only finitely many zeroes of N are thereby encountered. However, in light of Lemma 7.1 one can go further by shifting the contour in the inner integral in (7.1) from $\mathbb{R} + i\sigma$ to the curve $\Im(\zeta) = \max(\sigma, \varepsilon|\Re(\zeta)|^\gamma)$, where $\gamma = (m-2)/2(m-1)$; the proof of Lemma 7.1 yields a lower bound on N which ensures convergence, provided that ε is sufficiently small.

It would be desirable to say that E_σ becomes in some sense small as $\sigma \to +\infty$, and moreover satisfies better estimates than S_ν whenever $\sigma_\nu < \sigma$; we believe the latter to be false. As weak evidence in support of the former, we offer

Computation. *For each $0 < \sigma \notin \{\sigma_\nu\}$ there exists $C_\sigma < \infty$ such that for all k and all $y > 0$,*

$$|\frac{\partial^k}{\partial t^k} E_\sigma(z,t)| \le C_\sigma C_0^k \min\left\{(\sigma y)^{-mk-m-2}\Gamma(mk), |x|^{-mk-m-2}\Gamma(k)\right\}.$$

C_0 is an absolute constant. The gain is in the negative power of σ; the argument yields no control over C_σ. This estimate is really intended only for $t = 0$ and is extremely poor otherwise. Even when $t = 0$ it is not optimal, but we believe that it does accurately reflect the effect of increasing σ.

8. Non-analytic Szegö kernels in higher dimensions.

In $\mathbb{C}^{n+1}$ write $z = (z_0, z') \in \mathbb{C} \times \mathbb{C}^n$ and consider the hypersurface $M = \{\Im(z_0) = P(x)\}$ where $z' = x + iy \in \mathbb{R}^n + i\mathbb{R}^n$ and $P : \mathbb{R}^n \mapsto \mathbb{R}$ is a convex polynomial. Then the region above M is pseudoconvex.

Assume that $P(0) = 0$, that $\nabla P(0) = 0$ and that for certain $a_1, \ldots a_n > 0$, for all $x \in \mathbb{R}^n$ and $r > 0$,

$$P(r^{a_1} x_1, \ldots r^{a_n} x_n) \equiv rP(x).$$

Necessarily the a_j are reciprocals of integers. Assume that P is not strictly convex at the origin, that is, its Hessian is degenerate there. Assume finally that M is of finite type in the sense of D'Angelo; in the present special context, this means simply that P does not vanish identically along any line through the origin.

Proposition 8.1. *Under the above hypotheses, the Szegö kernel for M fails to be analytic at certain points off of the diagonal.*

Identify M with $\mathbb{C}^n \times \mathbb{R}$ by $(z, t) \mapsto (z, t + iP(x))$. Then by the Szegö kernel, we mean the distribution-kernel for the orthogonal projection of $L^2(\mathbb{C}^n \times \mathbb{R})$, with respect to Lebesgue measure, onto the intersection of the nullspaces of the n operators

$$\frac{\partial}{\partial \bar{z}_j} - i\frac{\partial P}{\partial \bar{z}_j}\frac{\partial}{\partial t}, \qquad 1 \le j \le n.$$

It may be computed explicitly, by taking the partial Fourier transform with respect to $t, y_1, \ldots y_n$ as was done for $n = 1$ in [N]. Writing $w = u + iv \in \mathbb{R}^n + i\mathbb{R}^n$ and defining

$$\delta_\tau(\eta) = (\tau^{a_1}\eta_1, \ldots \tau^{a_n}\eta_n),$$

one has

$$S((z,t),(w,s)) = (2\pi)^{-n-1} \int_0^\infty e^{i(t-s)\tau} e^{-\tau P(x) - \tau P(u)} \int_{\mathbb{R}^n} e^{\delta_\tau \eta \cdot (z + \bar{w})} N(\eta)^{-1} d\eta \, d\tau$$

where for $\zeta \in \mathbb{C}^n$,

$$N(\zeta) = \int_{\mathbb{R}^n} e^{2(\zeta \cdot s - P(s))} ds.$$

Arguing as was done in [CG] for $n = 1$, one finds that analyticity of $(y,t) \mapsto S((0 + iy, t), 0)$ away from the origin would preclude the existence of zeroes of N in $\mathbb{C}^n$. But since P fails to be strictly convex at the origin, after possibly renumbering the coordinates we have $P(x_1, 0, \ldots 0) = c_0 x_1^m$ where $m = a_1^{-1} > 2$. c_0 must be strictly positive, and m even, by the hypotheses of pseudoconvexity and finite type.

In fact, $N(\zeta_1, 0, \ldots, 0)$ is an entire function of order $m/(m-1)$ in $\mathbb{C}^1$. For,

$$N(\zeta_1, 0, \ldots) = c \int_0^\infty \int_{S^{n-1}} e^{\zeta_1 r \theta_1} e^{-r^m P(\theta)} d\theta \, r^\gamma \, dr$$

for a certain exponent γ and constant c. An upper bound for $\zeta_1 \in \mathbb{C}$ and lower bound for $\zeta \in \mathbb{R}$ follow, using the fact that P vanishes only at the origin. Therefore N has zeroes, and S is non-analytic.

REFERENCES

[BG] M. S. Baouendi and C. Goulaouic, *Nonanalytic-hypoellipticity for some degenerate elliptic operators*, Bull. AMS **78** (1972), 483-486.

[Ch] S. C. Chen, *Global analytic hypoellipticity of the $\bar{\partial}$-Neumann problem on circular domains*, Invent. Math. **92** (1988), 173-185.

[C1] M. Christ, *Embedding compact three-dimensional CR manifolds of finite type in $\mathbb{C}^n$*, Annals of Math. **129** (1989), 195-213.

[C2] _____, *On the $\bar{\partial}$ equation in $\mathbb{C}^1$ with weights*, preprint.

[C3] _____, *Some non-analytic-hypoelliptic sums of squares of vector fields*, preprint.

[C4] _____, *Analytic hypoellipticity breaks down for weakly pseudoconvex Reinhardt domains*, preprint.

[CG] M. Christ and D. Geller, *Counterexamples to analytic hypoellipticity for domains of finite type*, Annals of Math. (to appear).

[FK] C. L. Fefferman and J. J. Kohn, *Hölder estimates on domains of complex dimension two and on three dimensional CR manifolds*, Adv. Math. **69** (1988), 223-303.

[FS] G. B. Folland and E. M. Stein, *Estimates for the $\bar{\partial}_b$ complex and analysis on the Heisenberg group*, Comm. Pure Appl. Math. **27** (1974), 429-522.

[G] D. Geller, *Analytic Pseudodifferential Operators For The Heisenberg Group And Local Solvability*, Mathematical Notes 37, Princeton University Press, Princeton, NJ, 1990.

[GS] A. Grigis and J. Sjöstrand, *Front d'onde analytique et sommes de carrés de champs de vecteurs*, Duke Math. J. **52** (1985), 35-51.

[HH] N. Hanges and A. A. Himonas, *Singular solutions for sums of squares of vector fields*, preprint.

[He] B. Helffer, *Conditions nécessaires d'hypoanalyticité pour des opérateurs invariants à gauche homogènes sur un groupe nilpotent gradué*, Jour. Diff. Eq. **44** (1982), 460-481.

[H] L. Hörmander, *The Analysis of Linear Partial Differential Operators I*, Springer-Verlag, Berlin, 1983.

[K] J. J. Kohn, *Estimates for $\bar{\partial}_b$ on pseudoconvex CR manifolds*, Proc. Symp. Pure Math. **43** (1985), 207-217.

[N] A. Nagel, *Vector fields and nonisotropic metrics*, Beijing Lectures in Harmonic Analysis, E. M. Stein, ed., Princeton University Press, Princeton, NJ, 1986, pp. 241-306.

[PR] Pham The Lai and D. Robert, *Sur un problème aux valeurs propres non linéaire*, Israel J. Math **36** (1980), 169-186.

[Sj] J. Sjöstrand, *Singularités analytiques microlocales*, Astérisque **95** (1982).

[Sm] H. Smith, *A calculus for three-dimensional CR manifolds of finite type*, preprint.

[Tp] J-M. Trepeau, *Sur l'hypoellipticité analytique microlocale des opérateurs de type principal*, Comm. PDE **9** (1984), 1119-1146.

[Tv1] F. Trèves, *Analytic-hypoelliptic partial differential equations of principal type*, Comm. Pure Appl. Math. **24** (1971), 537-570.

[Tv2] _____, *Analytic hypo-ellipticity of a class of pseudodifferential operators with double characteristics and applications to the $\bar{\partial}$-Neumann problem*, Comm. PDE 13 (1978), 475-642.

DEPARTMENT OF MATHEMATICS, UCLA, LOS ANGELES, CA. 90024

E-mail: christ @ math.ucla.edu

L^2 Estimates in Nonlinear Fourier Analysis

R. Coifman and S. Semmes

1. Introduction

One of the goals of classical Fourier analysis is to obtain L^p estimates for linear operators that commute with translations. The case of $p = 2$ plays a special role, because of Placherel's theorem, and because of the availability of methods from the Calderón-Zygmund school for deriving L^p estimates from the L^2 case. However, the relevance of Plancherel's theorem fades swiftly to oblivion when we shift our attention to nonlinear objects. (The Calderón-Zygmund technology does not.) In this paper we present an approach to dealing with nonlinear functionals that commute with translations and which satisfy nonlinear versions of the Calderón-Zygmund conditions.

An example of the kind of nonlinear transformation we have in mind is the following. Let F be a smooth function on $\mathbf{R}$. Given real-valued functions a, f on $\mathbf{R}$, define $T(a, f)$ by

$$(1.1) \qquad T(a, f)(x) = P.V. \int_{\mathbf{R}} F\left(\frac{A(x) - A(y)}{x - y}\right) \frac{1}{x - y} f(y)\, dy,$$

where $A' = a$. For fixed a this can be viewed as a linear operator in f of nonconvolution type, but we prefer to think of it as a nonlinear functional in a and f that does commute with translations. That is, if we define τ_h by $\tau_h(f) = f(x - h)$, then

$$(1.2) \qquad \tau_h(T(a, f)) = T(\tau_h a, \tau_h f).$$

In [CDM] it is proved that

$$(1.3) \qquad \|T(a, f)\|_2 \leq C(F, \|a\|_\infty)\|f\|_2.$$

The argument relied heavily on the special form of $T(a, f)$; F was expressed as a linear superposition of translates and dilates of $\frac{1}{1+ix}$, which permitted reduction to known estimates for the Cauchy integral operator on Lipschitz graphs. The methods used to treat the Cauchy integral operator have never been successfully applied directly to $T(a, f)$ for general F.

The proof of our main result provides an approach for dealing with $T(a, f)$ as in (1.1), as well as any other operation that satisfies (1.2) and certain conditions of Calderón-Zygmund type. These conditions concern the dependence on both a and f, and in particular they take into account the nonlinearity. Roughly speaking they ensure that $T(a, f)$ has the same sort of localization properties in a and f as do linear Calderón-Zygmund operators.

In a moment we shall list the hypotheses that we impose on a general operation $T(\cdot, \cdot)$ for our main result. These hypotheses are not satisfied by the example given in (1.1), but we shall explain the modifications needed to rectify this after stating the theorem.

Both authors are supported by the National Science Foundation. The second author is also supported by the Alfred P. Sloan Foundation

For the sake of clarity we have tried to minimize the level of generality in the following discussion, subject only to the constraint of having (1.1) as an example. We have also included in our hypotheses many a priori assumptions, so that the theorem asserts the existence of estimates that do not depend on these assumptions in a quantitative way. In applications these a priori assumptions are not burdensome.

Fix $n \geq 1$. Let $C_0^\infty = C_0^\infty(\mathbf{R}^n)$ the space of C^∞ functions on $\mathbf{R}^n$ with compact support. Let $T : C_0^\infty \times C_0^\infty \to C_0^\infty$ be a continuous transformation that is linear in the second argument but not necessarily in the first. [Here and throughout terms such as "continuous" refer to the usual (strong) topology on C_0^∞.]

Actually, it will be helpful to allow the first functional argument in T to be the sum of a constant function and an element of C_0^∞. Let $C_1^\infty = C_1^\infty(\mathbf{R}^n)$ denote the space of functions of this form, and assume that T admits a continuous extension to $C_1^\infty \times C_0^\infty$ into C_0^∞.

The qualitative conditions we place on $T(\cdot, \cdot)$ are as follows. We assume that for each $a \in C_0^\infty$ there is a C^∞ function $K_a(x, y)$ on $\mathbf{R}^n \times \mathbf{R}^n$ such that

$$(1.4) \qquad T(a, f)(x) = \int_{\mathbf{R}^n} K_a(x, y) f(y) \, dy$$

for all $f \in C_0^\infty$, and that K_a depends continuously on a. We also impose similar requirements on the first and second differentials of T in a. Let us state these requirements slowly.

We demand that $T(a, f)$ be C^2 in a, in the following sense. Given any $\alpha, \beta, \gamma \in C_1^\infty$, we require that

$$(1.5) \qquad T'(a, \alpha, f) = \frac{d}{dt}\Big|_{t=0} (T(a + t\alpha, f)) \quad \text{and}$$

$$(1.6) \qquad T''(a, \beta, \gamma, f) = \frac{d}{ds}\Big|_{s=0} \frac{d}{dt}\Big|_{t=0} (T(a + s\beta + t\gamma, f))$$

exist in C_0^∞, and that they depend continuously on the functional parameters. We also demand that for each a there exist C^∞ functions $K_a'(x, y, z)$ and $K_a''(x, y, z, w)$ on $\mathbf{R}^n \times \mathbf{R}^n \times \mathbf{R}^n$ and $\mathbf{R}^n \times \mathbf{R}^n \times \mathbf{R}^n \times \mathbf{R}^n$ that depend continuously on a and satisfy

$$(1.7) \qquad T'(a, \alpha, f)(x) = \iint K_a'(x, y, z) f(y) \alpha(z) \, dy \, dz,$$

$$(1.8) \qquad T''(a, \beta, \gamma, f)(x) = \iiint K_a''(x, y, z, w) f(y) \beta(z) \gamma(w) \, dy \, dz \, dw$$

for all $a, \alpha, \beta, \gamma \in C_1^\infty$ and $f \in C_0^\infty$. [Here and in the future we suppress the range of integration when it is $\mathbf{R}^n$.]

For the quantitative hypotheses on T we assume that for each $A > 0$ there is a $B = B(A) > 0$ so that the following estimates hold when $\|a\|_\infty \leq A$.

We ask that the kernel K_a satisfy the usual standard estimates, i.e.,

$$(1.9) \qquad |K_a(x, y)| \leq B|x - y|^{-n},$$

(1.10)
$$|\nabla K_a(x,y)| \le B|x-y|^{-n-1}$$

for all $x,y \in \mathbf{R}^n$, where the gradient is taken in the x and y variables. We also make similar requests of K'_a, K''_a:

(1.11) $\quad |K'_a(x,y,z)| \le B(|x-y|+|x-z|)^{-2n},$

(1.12) $\quad |K''_a(x,y,z,w)| \le B(|x-y|+|x-z|+|x-w|)^{-3n},$

(1.13) $\quad |\nabla K'_a(x,y,z)| \le B(|x-y|+|x-z|)^{-2n-1},$ and

(1.14) $\quad |\nabla K''_a(x,y,z,w)| \le B(|x-y|+|x-z|+|x-w|)^{-3n-1}$

for all x,y,z,w, where the gradients in (1.13) and (1.14) are taken in all variables.

We further entreat these kernels to satisfy certain cancellation conditions, namely that whenever a is a $\underline{\text{constant}}$ function with $|a| \le A$ we have

(1.15)
$$\sup_{R>0} \left| \int_{B(x,R)} K_a(x,y)dy \right| \le B \quad \text{and}$$

(1.16O)
$$\sup_{R>0} \left| \int_{B(x,R)} \int_{B(x,R)} K'_a(x,y,z)dy\,dz \right| \le B$$

for all $x \in \mathbf{R}^n$, where $B(x,R)$ denotes the ball with center x and radius R. Similarly, if we define $\tilde{K}''_a(x,y,z)$ by

(1.17)
$$\tilde{K}''_a(x,y,z) = \int_{\mathbf{R}^n} K''_a(x,y,z,w)dw,$$

then we require that

(1.18)
$$\sup_{R>0} \left| \int_{B(x,R)} \int_{B(x,R)} \tilde{K}''_a(x,y,z)dy\,dz \right| \le B$$

for all $x \in \mathbf{R}^n$ and all constant functions a with $|a| \le A$.

THEOREM 1.19. *Suppose that $T(\cdot,\cdot)$ satisfies the conditions listed above, and that T commutes with translations, i.e., (1.2) holds. Then for every $A > 0$ there is a $B' = B'(A)$ so that*

(1.20)
$$\|T(a,f)\|_2 \le B'\|f\|_2$$

for all $a, f \in C_0^\infty$ with $\|a\|_\infty \le A$. The function $B'(\cdot)$ depends only on the dimension n and the function $B(\cdot)$.

For each value of A, $B'(A)$ depends only on n and $B(CA)$ for some universal constant C.

Of course L^p and other estimates can easily be obtained from (1.19) using Calderón-Zygmund theory.

The theorem as stated does not apply to the example given by (1.1). There are two reasons for this. The first is that the various qualitative a priori assumptions are not satisfied. This is easily fixed by performing standard truncations and regularizations. The second problem is more serious: the estimates (1.13) and (1.14) are not satisfied due to certain discontinuities in K'_a, K''_a. This can again be repaired by means that are standard, but, unfortunately, somewhat technical. We can replace (1.13) and (1.14) by more general integral conditions which are strong enough for the theorem to hold, but weak enough to be applicable.

When $n = 1$ there are integral conditions that do this and which are fairly simple. The substitute for (1.13) is

(1.21)

$$\int_{B(x,R)} |\nabla_u K'_a(u,y,z)| du + \int_{B(y,R)} |\nabla_u K'_a(x,u,z)| du$$

$$+ \int_{B(z,R)} |\nabla_u K'_a(x,y,u)| du \le BR^{-2}$$

for all x, y, z, where $R = \frac{1}{10}(|x - y| + |x - z|)$, and the substitute for (1.14) is that

(1.22)

$$\int_{B(x,S)} |\nabla_u K''_a(u,y,z,w)| du \le BS^{-3}$$

hold for all x, y, z, w, where $S = \frac{1}{10}(|x - y| + |x - z| + |x - w|)$, and that the analogous inequality where the role of x is played by y, z, or w also hold (as in (1.21)).

When $n > 1$ the obvious versions of (1.21) and (1.22) aren't strong enough for our proof of the theorem to work. We could strenghen (1.21) and (1.22) by using L^p norms, but this is too restrictive; we want to have conditions that allow jump discontinuities, at least on sufficiently small sets. A good but slightly complicated substitute for (1.13), (1.14) can be given as follows.

For each $R > 0$ define

$$D_3(R) = \{(x,y,z) : R/2 \le |x - y| + |x - z| < R\} \quad \text{and}$$
$$D_4(R) = \{(x,y,z,w) : R/2 \le |x - y| + |x - z| + |x - w| < R\},$$

and set

$$J'_R(x,y,z) = \chi_{D_3(R)}(x,y,z) |\nabla_x K'_a(x,y,z)|,$$
$$J''_R(x,y,z,w) = \chi_{D_4(R)}(x,y,z,w) |\nabla_x K''_a(x,y,z,w)|.$$

The substitutes for (1.13) and (1.14) are that there exist $p = p(A) \in [1, \infty]$ and $\delta = \delta(A) > 0$ such that

$$2\delta > n[2(\min(p,2))^{-1} - 1]$$

and such that the following hold for all $R, r > 0$ with $r \le R$:

(1.23) $$\left(R^{-n} \int_{\mathbb{R}^n} \{r^{-n} \int_{B(z,r)} R^{2n} J'_R(v,y,z) dv\}^p dx\right)^{\frac{1}{p}} \le Br^{\delta-1} R^{-\delta}$$

for all $y, z \in \mathbf{R}^n$, and similarly when each of y and z plays the role of x in (1.23) and the definition of $J'_R(x, y, z)$; and

$$(1.24) \qquad (R^{-n} \int_{\mathbf{R}^n} \{r^{-n} \int_{B(x,r)} R^{3n} J''_R(v, y, z, w) dv\}^P dx)^{\frac{1}{p}} \leq Br^{\delta-1} R^{-\delta}$$

for all $y, z, w \in \mathbf{R}^n$, and similarly when each of y, z, and w plays the role of x in (1.24) and the definition of $J''_R(x, y, z, w)$.

When we prove the theorem we shall show that it is enough to assume these conditions instead of (1.13) and (1.14). Notice that (1.13) and (1.14) correspond to the case of $p = \infty$, $\delta = 1$ in (1.23) and (1.24), while (1.21) and (1.22) correspond to $p = 1$, $\delta = 1$.

It is not hard to check that the example in (1.1) can be truncated and regularized in such a way that the hypothesis of Theorem 1.19 hold with uniform bounds, except that (1.13) and (1.14) are replaced by (1.21) and (1.22). However, it is a little easier to treat this example directly with the methods of the proof than it is to do the general case.

2. The Proof of the Theorem

The proof will be effected through localizations in space and frequency that allow square function estimates to be used, as in [CM].

Let φ be a C^∞ function on $\mathbf{R}^n$ that is even, supported in $B(0, 1)$, and which satisfies

$$(2.1) \qquad \int \varphi = 1, \quad \int \varphi(x) p(x) dx = 0$$

for all polynomials $p(x)$ of degree at most 10 with $p(0) = 0$. Define the operator P_t by $P_t f = \varphi_t * f$, where $\varphi_t(x) = t^{-n} \varphi(\frac{x}{t})$. Define Q_t by $Q_t = t \frac{\partial}{\partial t} P_t$, so that $Q_t f = \psi_t * f$ for a function $\psi \in C_0^\infty$ that is supported in $B(0, 1)$. It is not hard to check that

$$(2.2) \qquad \int \psi(x) p(x) dx = 0$$

for all polynomials of degree ≤ 10, including constants this time.

Fix $a, f, g \in C_0^\infty$ with $\|a\|_\infty \leq A$. We want to show that

$$\left| \int T(a, f) g \right| \leq C \|f\|_2 \|g\|_2$$

with a constant C that depends only on A, n, and the function $B(\cdot)$ in the hypotheses of Theorem 1.19. Throughout this section we let C denote any constant that depends only on these things.

We shall employ the usual trick of writing

$$(2.3) \qquad \int_{\mathbf{R}^n} T(a, f) g dx = \int_0^\infty \int_{\mathbf{R}^n} t \frac{\partial}{\partial t} \{T(P_t a, P_t f) P_t g\} dx \frac{dt}{t}.$$

This is a sum of three pieces, $I + II + III$, where

(2.4)
$$I = \int_0^\infty \int_{\mathbf{R}^n} T'(P_t a, Q_t a, P_t f) P_t g \, dx \frac{dt}{t},$$

(2.5)
$$II = \int_0^\infty \int_{\mathbf{R}^n} T(P_t a, Q_t f) P_t g \, dx \frac{dt}{t}, \text{ and}$$

(2.6)
$$III = \int_0^\infty \int_{\mathbf{R}^n} T(P_t a, P_t f) Q_t g \, dx \frac{dt}{t}.$$

The third term is essentially the transpose of the second. That is, if you replace $T(\cdot, \cdot)$ in II by its transpose (as a linear operator in the second argument), then you get III. Hence III can be treated in the same way as II, and so we need only consider I and II.

In order to control I and II by square functions we shall integrate by parts and make other manipulations so as to produce a second Q_t in each term. For this endeavor the assumption of translation invariance is crucial. Before doing this we review briefly some well-known facts about quadratic estimates, mostly for the purpose of setting notations.

Let us first recall Carleson's inequality. Let CM ("Carleson measures") denote the space of functions $M(x, t)$ on $\mathbf{R}_+^{n+1}$ such that

$$t^{-n} \int_0^t \int_{B(x,t)} M(y, s) dy \frac{ds}{s}$$

is uniformly bounded for $x \in \mathbf{R}^n$, $t > 0$, and let $\|M\|_{CM}$ denote the least upper bound. Let QCM ("quadratic Carleson measures") denote the space of functions M on $\mathbf{R}^{n+1}$ such that $M^2 \in CM$, and set $\|M\|_{QCM} = \|M^2\|_{CM}^{\frac{1}{2}}$. Carleson's inequality says that

(2.7)
$$\int_0^\infty \int_{\mathbf{R}^n} F(x, t) M(x, t) dx \frac{dt}{t} \le C \|M\|_{CM} \|F^*\|_{L^1(\mathbf{R}^n)},$$

where $F^*(x) = \sup\{|F(y, t)|; |y - x| \le t\}$ is the nontangential maximal function of F.

Next we recall the basic square function and Carleson measure estimates. Let η be a function on $\mathbf{R}^n$ such that

$$|\eta(x)| \le C(1 + |x|)^{-n-1}, \quad |\nabla \eta(x)| \le C(1 + |x|)^{-n-1}, \text{ and } \int \eta(x) dx = 0.$$

Then

(2.8)
$$\int_0^\infty \int_{\mathbf{R}^n} |\eta_t * h(x)|^2 dx \frac{dt}{t} \le C \|h\|_2^2$$

for all $h \in L^2(\mathbf{R}^n)$, while

(2.9)
$$\|\eta_t * h(x)\|_{QCM} \le C \|h\|_\infty$$

whenever $h \in L^\infty(\mathbf{R}^n)$. The first inequality can be derived from Plancherel, while the second follows from the first and a standard localization argument. See [CM] or [JL], for instance.

There is a slight variation of this that we shall need. Suppose that

$$|\nabla^j \eta(x)| \le C(1 + |x|)^{-n-1}$$

for $j = 1, 2, \ldots, N$, $N = [n/2] + 2$. Then

(2.10)
$$\int_0^\infty \int_{\mathbf{R}^n} \{\eta_t * h\}_{x,t}^2 \, dx \frac{dt}{t} \le C\|h\|_2^2,$$

where

(2.11)
$$\{G\}_{x,t} = \sup_{y \in B(x,t)} (|G(y)| + t|\nabla G(y)|).$$

Indeed, it is not hard to show that

$$\{\eta_t * h\}_{x,t}^2 \le Ct^{-n} \int_{B(x,2t)} \sum_{j=0}^N (t^j |\nabla^j \eta_t * h(y)|)^2 \, dy,$$

and this permits (2.10) to be reduced to (2.8) using Fubini's theorem. [This proof is horribly crude, because $N = 2$ is more than enough for (2.10) to hold, but it is simple and adequate for our purposes.] Similarly,

(2.12)
$$\|\{\eta_t * h\}_{x,t}\|_{QCM} \le C\|h\|_\infty.$$

Before beginning the analysis of I and II there is a little more notation that we should introduce now. Let $\theta(x)$ be a C^∞ function on $\mathbf{R}^n$ that is supported in $B(0,1)$ and which satisfies $0 \le \theta \le 1$ on $\mathbf{R}^n$ and $\theta \cdot = 1$ on $B(0, \frac{1}{2})$. Set $\theta(x, t) = \theta(t^{-1}x)$ and

$$\theta(x, y, z, t) = \theta(x - y, t)\theta(x - z, t).$$

Also, we shall let dV_2, dV_3, and dV_4 represent $dx\,dy\frac{dt}{t}$, $dx\,dy\,dz\frac{dt}{t}$, and $dx\,dy\,dz\,dw\frac{dt}{t}$.

To analyze I we split it into $Ia + Ib$, where

(2.13)
$$Ia = \int_0^\infty \iiint \theta(x, y, z, t) K'_{P_t a}(x, y, z) Q_t a(z) P_t f(y) P_t g(x) dV_3,$$

and where Ib is given by the same expression but with $1 - \theta(x, y, z, t)$ instead of $\theta(x, y, z, t)$. To control Ia we split it further into two terms $Ia(i)$ and $Ia(ii)$, according to the decomposition

(2.14)
$$K'_{P_t a}(x, y, z) = (K'_{P_t a}(x, y, z) - K'_{P_t a(x)}(x, y, z)) + K'_{P_t a(x)}(x, y, z).$$

To prevent any possible confusion we should state very explicitly what $K_{P_t a(x)}(x, y, z)$ means. Given x, y, z, and t we compute $P_t a(x)$ and get a number. We can also view this number as a function, and use it as the functional parameter in K' to get $K'_{P_t a(x)}(\cdot, \cdot, \cdot)$. We then evaluate this function at (x, y, z) to get $K'_{P_t a(x)}(x, y, z)$.

To estimate $Ia(i)$ we use the following inequality:

(2.15)
$$|K'_{P_t a}(x, y, z) - K'_{P_t a(x)}(x, y, z)|$$
$$\leq C \int_{\mathbb{R}^n} (|x - y| + |x - z| + |x - w|)^{-3n} |P_t a(x) - P_t a(w)| dw.$$

This follows from (1.12) by integrating
(2.16)
$$\frac{\partial}{\partial s} K'_{s P_t a + (1-s) P_t a(x)}(x, y, z) = \int K''_{s P_t a + (1-s) P_t a(x)}(x, y, z, w)(P_t a(w) - P_t a(x)) dw$$

in s, and it implies that

(2.17)
$$|Ia(i)| \leq C \int_0^\infty \iiint \theta(x, y, z, t) H(x, y, z, t) |Q_t a(z)| |P_t f(y)| |P_t g(x)| dV_3$$

where $H(x, y, z, t)$ denotes the right side of (2.15). Let $H_1(x, y, z, t)$ and $H_2(x, y, z, t)$ be the parts of $H(x, y, z, t)$ that correspond to $|x - w| \leq t$ and $|x - w| \geq t$, respectively, and let $Ia(i)1$ and $Ia(i)2$ denote the corresponding pieces of the right side of (2.17).

Because
$$|P_t a(w) - P_t a(x)| \leq t^{-1} |x - w| \{t|\nabla P_t a|\}_{x,t}$$

when $|x - w| \leq t$, by the mean-value theorem, we have that

$$H_1(x, y, z, t) \leq C t^{-1} (|x - y| + |x - z|)^{-2n+1} \{t|\nabla P_t a|\}_{x,t}$$
$$\leq C(t^{-\frac{1}{2}} |x - y|^{-n+\frac{1}{2}})(t^{-\frac{1}{2}} |x - z|^{-n+\frac{1}{2}}) \{t|\nabla P_t a|\}_{x,t}.$$

Inserting this into $Ia(i)1$ and integrating in y and z we get that

$$Ia(i)1 \leq C \int_0^\infty \int_{\mathbb{R}^n} \{t|\nabla P_t a|\}_{x,t} N(x, t) F(x, t) |P_t g(x)| \frac{dx \, dt}{t},$$

where

$$N(x, t) = \int_{B(x,t)} t^{-\frac{1}{2}} |x - z|^{-n+\frac{1}{2}} |Q_t a(z)| dz,$$

$$F(x, t) = \int_{B(x,t)} t^{-\frac{1}{2}} |x - y|^{-n+\frac{1}{2}} |P_t f(y)| dy.$$

Clearly $N(x, t) \leq C\{|Q_t a|\}_{x,t}$, and so

$$\|\{t|\nabla P_t a|\}_{x,t} N(x, t)\|_{CM} \leq C,$$

by (2.12). Since the nontangential maximal functions of $F(x,t)$ and $|P_t g(x)|$ have L^2 norms dominated by $\|f\|_2$ and $\|g\|_2$, we get that

$$Ia(i)1 \leq C\|f\|_2\|g\|_2.$$

To control $Ia(i)2$ we begin with a trivial upper bound for H_2, namely

$$H_2(x,y,z,t) \leq \int_{|x-w|\geq t} |x-w|^{-3n}|P_t a(w) - P_t a(x)|dx.$$

Let $M(x,t)$ denote t^{2n} times the right side, so that

$$Ia(i)2 \leq C \int_0^\infty \int_{\mathbf{R}^n} M(x,t)N(x,t)F(x,t)|P_t g(x)|dx\frac{dt}{t},$$

where we can now take $N(x,t) = t^{-n} \int_{B(x,t)} |Q_t a(z)|dz$ and $F(x,t) = t^{-n} \int_{B(x,t)} |P_t f(y)|dy$.
If we can show that $\|M\|_{QCM} \leq C$, then we have $Ia(i)2 \leq C\|f\|_2\|g\|_2$, just as before.
To control M we use

$$M(x,t) \leq \int_{|x-w|\geq t} t^{2n}|x-w|^{-3n+1} \int_0^1 |\nabla P_t a(x + s(w-x))|dsdw.$$

Next we interchange the order of integration and make the change of variables $w \mapsto u = x + s(w - x)$ to get

$$M(x,t) \leq \int_0^1 \int_{|x-u|\geq st} t^{2n}s^{3n-1}|x-u|^{-3n+1}|\nabla P_t a(u)|s^{-n}duds$$

$$= \int_{\mathbf{R}^n} t^{2n}\left(\int_0^{\min(1,t^{-1}|x-u|)} s^{2n-1}ds\right)|x-u|^{-3n+1}|\nabla P_t a(u)|du$$

$$\leq C\int_{\mathbf{R}^n} t^{2n}\min(1,t^{-1}|x-u|)^{2n}|x-u|^{-3n+1}|\nabla P_t a(u)|du.$$

Using this it is not hard to show that $\|M\|_{QCM} \leq C$, using the corresponding result for $t|\nabla P_t a(x)|$ and $\{t|\nabla P_t a|\}_{x,t}$. (It is helpful to consider separately the $|x - u| \geq t$ and $|x - u| \leq t$ parts of the integral.)
This completes the proof of

$$|Ia(i)| \leq C\|f\|_2\|g\|_2,$$

and so now we consider $Ia(ii)$, which is given by

(2.18) $$\int_0^\infty \iiint \theta(x,y,z,t)K'_{P_t a(x)}(x,y,z)Q_t a(z)P_t f(y)P_t g(x)dV_3.$$

To bound this using quadratic estimates we need to have another Q_t in the integral. We shall get it by an integration by parts, by pulling some derivatives out of the $Q_t a(z)$, after introducing some more notation and performing a preliminary calculation.

Set $\tilde{\psi} = \Delta^{-1}\psi$ and $\tilde{\psi}^j = \partial_j\tilde{\psi}$, where ∂_j denotes differentiation in the jth co-ordinate. Thus $\tilde{\psi}$ and $\tilde{\psi}^j$ have the same properties that ψ does, except that they don't have quite as many vanishing moments, and instead of being compactly supported they have fairly rapid decay at infinity (like $|x|^{-n-9}$ and $|x|^{-n-10}$, respectively). Let $\tilde{Q}_t$ and $\tilde{Q}_t^j$ denote the operators of convolution by $\tilde{\psi}_t$ and $\tilde{\psi}_t^j$, and set $Q_t^j = t\,\partial_j P_t$.

Thus we have

(2.19)
$$Q_t a = \sum_{j=1}^{n} t\partial_j \tilde{Q}_t^j a.$$

For the integration by parts that we are about to do it turns out to be better to use a slightly different identity, to wit,

(2.20)
$$Q_t a(z) = \sum_{j=1}^{n} t\mathcal{D}_j(\tilde{Q}_t^j a(z)),$$

where $\mathcal{D}_j = \frac{\partial}{\partial x_j} + \frac{\partial}{\partial y_j} + \frac{\partial}{\partial z_j}$.

The reason that this is better is that $\mathcal{D}_j(\theta(x,y,z,t)) = 0$ and

(2.21)
$$\mathcal{D}_j(K'_\alpha(x,y,z)) = 0$$

for all constant functions α. This last equation is a consequence of (1.2), because (1.2) implies that

(2.22)
$$\tau_h(T'(\cdot,\cdot,\cdot)) = T'(\tau_h(\cdot), \tau_h(\cdot), \tau_h(\cdot)),$$

and hence that

(2.23)
$$K'_a(x - h, y, z) = K'_{\tau_h(a)}(x, y + h, z + h).$$

We are not going to use (2.21) itself, but rather its consequence that

$$tD_j(K'_{P_t\,a(z)}(x,y,z)) = t\frac{\partial}{\partial u_j}(K'_{P_t\,a(u)}(x,y,z))\big|_{u=x}$$
$$= \left[\int_{\mathbf{R}^n} K''_{P_t\,a(u)}(x,y,z,w)t\frac{\partial}{\partial u_j}(P_t a(u))dw\right]_{u=x}$$
$$= \tilde{K}''_{P_t\,a(z)}(x,y,z)Q_t^j a(x),$$

where $\tilde{K}''$ is as in (1.17).

We are now ready to do the integration by parts in (2.18). We replace $Q_t a(z)$ in (2.18) by the right side of (2.20), and then integrate the $\mathcal{D}_j$ by parts. The result can be written as $-S_1 - S_2$, where

S_1

(2.25)

$$= \sum_j \int_0^\infty \iiint \theta(x,y,z,t)\tilde{K}''_{P_t a(x)}(x,y,z)Q_t^j a(x)\tilde{Q}_t^j a(z)P_t f(y)P_t g(x)dV_3,$$

S_2

$$= \sum_j \int_0^\infty \iiint \theta(x,y,z,t)K'_{P_t a(x)}(x,y,z)\tilde{Q}_t^j a(z)\{Q_t^j f(y)P_t g(x) + P_t f(y)Q_t^j g(x)\}dV_3.$$

These two expressions will be treated in approximately the same manner.

Let $\hat{S}_1$ be the quantity you get by replacing $\tilde{Q}_t^j a(z)$ and $P_t f(y)$ in (2.25) by $\tilde{Q}_t^j a(x)$ and $P_t f(x)$, and let $\hat{S}_2$ be the quantity that results from substituting $\tilde{Q}_t^j a(x), Q_t^j f(x)$, and $P_t f(x)$ for $\tilde{Q}_t^j a(z), Q_t^j f(y)$, and $P_t f(y)$ in (2.26). Let us show that

(2.27)
$$\sum_{i=1}^2 |\hat{S}_i| + |S_i - \hat{S}_i| \le C\|f\|_2\|g\|_2.$$

Observe that

$$\left| \iint \theta(x,y,z,t)\tilde{K}''_{P_t a(x)}(x,y,z)dydz\right| \le C,$$

because of (1.18) and (1.12). Thus

$$|\hat{S}_1| \le C\sum_j \int_0^\infty \int_{\mathbf{R}^n} |Q_t^j a(x)||\tilde{Q}_t^j a(x)||P_t f(x)||P_t g(x)|dx\frac{dt}{t},$$

and this is bounded by $C\|f\|_2\|g\|_2$, because of Carleson's inequality and (2.9). The estimation of $|\hat{S}_2|$ is similar, although (2.8) and Cauchy-Schwarz enter the fray at the last step.

The point behind the estimate of $|S_1 - \hat{S}_1|$ is that

$$|\tilde{Q}_t^j a(z) - \tilde{Q}_t^j a(x)| \le \frac{|x-z|}{t}\{\tilde{Q}_t^j a\}_{x,t}$$

when $|x-z| \le t$, and similarly for $|P_t f(y) - P_t f(x)|$. This and (1.12) allow us to dominate $|S_1 - \hat{S}_1|$ by

$$C\sum_j \int_0^\infty \int I(x,t)|Q_t^j a(x)|\{\tilde{Q}_t^j a\}_{x,t}\{P_t f\}_{x,t}|P_t g(x)|dx\frac{dt}{t},$$

where $\tilde{I}(x,t) = \iint t^{-1}\theta(x,y,z,t)(|x-y|+|x-z|)^{-2n+1}dydz$. Clearly $\tilde{I}(x,t) \le C$ and so $|S_1 - \hat{S}_1| \le C\|f\|_2\|g\|_2$ can be derived from Carleson's inequality again. The estimation of $|S_2 - \hat{S}_2|$ is similar.

This proves (2.27), and hence that $|Ia(ii)| \le C\|f\|_2\|g\|_2$. Combining this with our previous bound for $Ia(i)$, we get the same kind of inequality for Ia.

Next we consider Ib, which is given by

$$(2.28) \qquad \int_0^\infty \iiint (1 - \theta(x,y,z,t))K'_{P_t a}(x,y,z)Q_t a(z)P_t f(y)P_t g(x)dV_3.$$

Our first task is to integrate by parts to get a second Q_t. This time we use the following variation of (2.24), which can be derived from (2.23):

$$(2.29) \qquad tD_j(K'_{P_t a}(x,y,z)) = \int K''_{P_t a}(x,y,z,w)Q_t^j a(w)dw.$$

Replace the $Q_t a(z)$ in (2.28) by the right side of (2.20), and integrate by parts. This yields $Ib = -\sum_j U_j - \sum_j V_j$, where

$$(2.30)$$
$$U_j = \int_0^\infty \iiiint (1 - \theta(x,y,z,t))K''_{P_t a}(x,y,z,w)Q_t^j a(w)\tilde{Q}_t^j a(z)P_t f(y)P_t g(x)dV_4,$$

$$(2.31)$$
$$V_j = \int_0^\infty \iiint (1 - \theta(x,y,z,t))K'_{P_t a}(x,y,z)\tilde{Q}_t^j a(z)\{Q_t^j f(y)P_t g(x) + P_t f(y)Q_t^j g(x)\}dV_3.$$

Although we now have two Q_t's in each term, we cannot yet apply the usual quadratic estimates, because the kernels K', K'' do not have adequate decay for this. Instead we first write $\tilde{Q}_t^j a(z)$ as $t\frac{\partial}{\partial z_j}\tilde{Q}_t a(z)$ and integrate by parts in z. This produces two terms in each U_j and V_j, according to whether the $\frac{\partial}{\partial z_j}$ hits the θ or the K'' or K'. We denote these terms by $U_{1,j}, V_{1,j}, U_{2,j}$, and $V_{2,j}$, where the subscript 1 corresponds to the $\frac{\partial}{\partial z_j}$ hitting the θ.

Consider $U_{1,j}$, which is given by

$$U_{i,j} = -\int_0^\infty \iiiint t\frac{\partial}{\partial z_j}(\theta(x,y,z,t))K''_{P_t a}(x,y,z,w)Q_t^j a(w)\tilde{Q}_t a(z)P_t f(y)P_t g(x)dV_4.$$

Clearly $t\frac{\partial}{\partial z_j}(\theta(x,y,z,t))$ is bounded, and it is nonzero only when $|x-y| \le t$, $|x-z| \le t$, and $|x-z| \ge \frac{1}{2}t$. When x,y,z, and t satisfy these inequalities we have

$$|K''_{P_t a}(x,y,z,w)| \le C(t + |x-w|)^{-3n}$$

because of (1.12). Hence

$$|U_{1,j}| \le \int_0^\infty \int_{\mathbf{R}^n} M(x,t)N(x,t), F(x,t)|P_t g(x)|dx\frac{dt}{t},$$

where $M(x,t) = \int_{\mathbf{R}^n} t^{2n}(t + |x - w|)^{-3n}|\tilde{Q}_t a(w)|dw$, and where $N(x,t)$ and $F(x,t)$ are the averages of $|\tilde{Q}_t a(z)|$ and $|P_t f(y)|$ over $B(x,t)$. It is easy to bound $|U_{1,j}|$ by $C\|f\|_2\|g\|_2$ using Carleson's inequality, because $\|M\|_{QCM} + \|N\|_{QCM} \leq C$. Similar reasoning produces the same bound for $|V_{1,j}|$.

The idea behind the estimates for $U_{2,j}$ and $V_{2,j}$ is simply that $t\frac{\partial}{\partial z_j}K'$ and $t\frac{\partial}{\partial z_j}K''$ have adequate decay to successfully reduce to square function estimates and Carleson's inequality. The implementation of this idea is somewhat complicated by the fact that we do not want to use (1.13) and (1.14), but instead the weaker conditions (1.23) and (1.24).

Let $D_3(R)$ and $D_4(R)$ be as in Section 1. [They are defined just before (1.23).] Then $(1 - \theta(x,y,z,t)) = 0$ when $(x,y,z,w) \in D_4(2^\ell t)$, and $\ell < 0$, and so

$$(2.32) \qquad |U_{2,j}| \leq \sum_{\ell=0}^{\infty} \int_0^\infty \iint H''_{2^\ell t}(x,y,w,t)|Q_t^j a(w)||P_t f(y)||P_t g(x)|dxdydw\frac{dt}{t},$$

where

$$H''_R(x,y,w,t) = \int tJ''_R(x,y,z,w)|\tilde{Q}_t a(z)|dz,$$

and $J''_R(x,y,z,w) = \chi_{D_4(R)}(x,y,z,w)|\nabla_z K(x,y,z,w)|$. Because $H''_R(x,y,w,t) = 0$ unless $|x - y| + |x - w| \leq R$, we have

$$(2.33) \qquad |U_{2,j}| \leq C\sum_{\ell=0}^{\infty} \int_0^\infty \int_{\mathbf{R}^n} M_\ell(x,t)N_\ell(x,t)F_\ell(x,t)|P_t g(x)|dx\frac{dt}{t},$$

where

$$M_\ell(x,t) = (2^\ell t)^{-n}\int_{B(x,2^\ell t)}|Q_t^j a(w)|dw,$$

$$N_\ell(x,t) = \sup_{y,w}(2^\ell t)^{2n}H''_{2^\ell t}(x,y,w,t),$$

$$F_\ell(x,t) = (2^\ell t)^{-n}\int_{B(x,2^\ell t)}|P_t f(y)|dy.$$

We are going to control (2.33) using Carleson's inequality.

It is not hard to check that the L^2 norm of the nontangential maximal function of F_ℓ is bounded by $C\|f\|_2$, uniformly in ℓ. Hence

$$(2.34) \qquad |U_{2,j}| \leq C\left(\sum_{\ell=0}^{\infty}\|M_\ell\|_{QCM}\|N_\ell\|_{QCM}\right)\|f\|_2\|g\|_2.$$

Next, let us verify that

$$(2.35) \qquad \|M_\ell\|_{QCM} \leq C\ell^{\frac{1}{2}}.$$

Fix x_0, t_0, and consider

(2.36)
$$\int_0^{t_0} \int_{B(x_0, t_0)} M_\ell(x, t)^2 \, dx \frac{dt}{t}.$$

When $t \geq 2^{-\ell} t_0$ we use the trivial estimate $M_\ell(x, t) \leq C$, so that the contribution to (2.36) is bounded by $C \ell t_0^n$. On the other hand,

$$\int_0^{2^{-\ell} t_0} \int_{B(x_0, t_0)} M_\ell(x, t)^2 \, dx \frac{dt}{t}$$

$$\leq C \int_0^{2^{-\ell} t_0} \int_{B(x_0, 2t_0)} |Q_t^j a(w)|^2 \, dw \frac{dt}{t} \leq C t_0^n.$$

This proves (2.36)

The argument for $\|N_\ell\|_{QCM}$ is more complicated but similar in spirit. We want to control $N_\ell(x, t)$ by some sort of average of $|\tilde{Q}_t a(z)|$. If we had (1.14) we could do exactly that, but with only (1.24) we shall control $N_\ell(x, t)$ by an L^q-average of $\{\tilde{Q}_t a\}_{z,t}$ for a suitable q.

We first want to modify H_R'' a little, so that K'' gets smoothed out. [This could be done in a more subtle manner, by not bringing in the absolute value signs as fast as we did, to allow (1.23) and (1.24) to be weakened.] It is not hard to check (using Fubini) that

(2.37)
$$H_R''(x, y, w, t) \leq C \int (t^{-n} \int_{B(z, t)} t J_R''(x, y, u, w) du) \{\tilde{Q}_t a\}_{z,t} dz.$$

Now we apply (1.24), with $r = t$ (which we assume is $\leq R$), and with z playing the role that x had. This gives

(2.38)
$$\left(R^{-1} \int_{\mathbb{R}^n} \{t^{-n} \int_{B(z, t)} R^{3n} J_R''(x, y, u, w) du\}^p dz\right)^{\frac{1}{p}}$$
$$\leq C r^{\delta - 1} R^{-\delta},$$

for all x, y and w, where p and δ are as in (1.24). [To be honest, p and δ are not $p(A), \delta(A)$, but rather $p(CA), \delta(CA)$, where C is chosen so that $\|P_t(a)\|_\infty \leq C \|a\|_\infty$.]

The conditions on p and δ are such that we may as well assume that $1 \leq p \leq 2$. Let q denote the conjugate index of p. For convenience of presentation we assume also that $q < \infty$, but the modifications needed for $q = \infty$ are completely trivial. From (2.37), (2.38), and Hölder's inequality we get that

$$H_R''(x, y, w, t) \leq C t R^{-2n} t^{\delta - 1} R^{-\delta} (R^{-n} \int_{B(x, 2R)} \{\tilde{Q}_t a\}_{z,t}^q dz)^{\frac{1}{q}}.$$

We have also used the fact that $J_R''(x, y, u, w) = 0$ unless $|x - u| \leq R$.

Taking $R = 2^\ell t, \ell \geq 0$, we obtain

$$(2.39) \qquad N_\ell(x,t) \leq C2^{-\delta\ell}\left((2^\ell t)^{-n}\int_{B(x,2^{\ell+1}t)}\{\tilde{Q}_t a\}^q_{z,t}dz\right)^{\frac{1}{q}}.$$

To estimate $\|N_\ell\|_{QCM}$ we want to convert the L^q norm to an L^2 norm. We can do this, but we will have to pay for it with a constant that depends on ℓ.

It is easy to check that

$$\sup_{z\in B(x,2^{\ell+1}t)}\{\tilde{Q}_t a\}_{z,t} \leq C\left(t^{-n}\int_{B(x,2^{\ell+1}t)}\{\tilde{Q}_t a\}^2_{t,2t}dz\right)^{\frac{1}{2}},$$

since $\{\tilde{Q}_t a\}_{z,t} \leq C\{\tilde{Q}_t a\}_{\xi,2t}$ whenever $|z-\xi| \leq t$. Hence

$$(2.40)$$

$$N_\ell(x,t) \leq C2^{-\delta\ell}\sup_{z\in B(x,2^{\ell+1}t)}\{\tilde{Q}_t a\}^{1-\frac{2}{q}}_{z,t}\left((2^\ell t)^{-n}\int_{B(x,2^{\ell+1}t)}\{\tilde{Q}_t a\}^2_{z,t}dz\right)^{\frac{1}{q}}$$

$$\leq C2^{-\delta\ell}2^{n\ell(\frac{1}{2}-\frac{1}{q})}\left((2^\ell t)^{-n}\int_{B(x,2^{\ell+1}t)}\{\tilde{Q}_t a\}^2_{z,2t}dz\right)^{\frac{1}{2}}$$

Let us now estimate $\|N_\ell\|_{QCM}$. Fix x_0, t_0, and consider

$$(2.41) \qquad \int_0^{t_0}\int_{B(x_0,t_0)}N_\ell(x,t)^2\,dx\frac{dt}{t}$$

The contribution of the $t \geq 2^{-\ell}t_0$ part of (2.41) is bounded by $C\ell 2^{-2\delta\ell}t_0^n$, because $N_\ell(x,t) \leq C2^{-\delta\ell}$, by (2.39). The remaining piece can be controlled by (2.40) to give

$$\int_0^{2^{-\ell}t_0}\int_{B(x_0,t_0)}N_\ell(x,t)^2\,dx\frac{dt}{t} \leq C\int_0^{2^{-\ell}t_0}\int_{B(x_0,3t_0)}2^{-2\delta}2^{2n\ell(\frac{1}{2}-\frac{1}{q})}\{\tilde{Q}_t a\}^2_{z,2t}dz\frac{dt}{t}$$

$$\leq C2^{-2\delta}2^{2n\ell(\frac{1}{2}-\frac{1}{q})}t_0^n.$$

Hence $\|N_\ell\|_{QCM} \leq C(\ell^{\frac{1}{2}} + 2^{n\ell(\frac{1}{2}-\frac{1}{q})})2^{-\delta\ell}$.

The constraints that were put on δ, p in Section 1 imply that $\|N_\ell\|_{QCM} \leq C2^{-\varepsilon\ell}$ for some $\varepsilon > 0$. This together with (2.34) and (2.35) give us that $|U_{2,j}| \leq C\|f\|_2\|g\|_2$.

Essentially the same argument gives $|V_{2,j}| \leq C\|f\|_2\|g\|_2$ also. This and our earlier estimates for $U_{1,j}$ and $V_{1,j}$ imply the same bound for Ib. Since we had already taken care of Ia, we have that $|I| \leq C\|f\|_2\|g\|_2$.

We can write II as

$$(2.42) \qquad II = \int_0^\infty\iint K_{P_t a}(x,y)Q_t f(y)P_t g(x)dV_2.$$

It turns out that we have to deal with II a bit differently than I, because the estimate for the analogue of $Ia(i)2$ for II does not work, due to inadequate decay at infinity. Instead we use an alternate approach in which we integrate by parts at the very beginning. This approach could have been applied to I, but at the cost of requiring control on the kernel of the third differential of T.

Using (2.20), with a replaced by f, we can integrate by parts in (2.42) to obtain

$$(2.43) \qquad II = -\sum_j \int_0^\infty \iiint K'_{P_t a}(x,y,z) Q_t^j a(z) \tilde{Q}_t^j f(y) P_t g(x) dV_3$$

$$-\sum_j \int_0^\infty \iint K_{P_t a}(x,y) \tilde{Q}_t^j f(y) Q_t^j g(x) dV_2.$$

We have also used here the analogue of (2.29) for K instead of K'.

Call these two sums IIa and IIb. We split them further by inserting $\theta(x,y,z,t)$ and $(1 - \theta(x,y,z,t))$ into the integrals of IIa to get $IIa(i)$ and $IIa(ii)$, and by inserting $\theta(x-y,t)$ and $(1 - \theta(x-y,t))$ into the integrals of IIb to get $IIb(i)$ and $IIb(ii)$.

Consider $IIb(i)$ first, which is given by

$$(2.44) \qquad -\sum_j \int_0^\infty \iint \theta(x-y,t) K_{P_t a}(x,y) \tilde{Q}_t^j f(y) Q_t^j g(x) dV_2.$$

Replace $K_{P_t a}(x,y)$ by

$$(K_{P_t a}(x,y) - K_{P_t a(x)}(x,y)) + K_{P_t a(x)}(x,y),$$

just as we did for Ia, to get $IIb(i)1$ and $IIb(i)2$.

Just as in (2.15) we have

$$(2.45) \quad |K_{P_t a}(x,y) - K_{P_t a(x)}(x,y)| \le C \int_{\mathbb{R}^n} (|x-y| + |x-z|)^{-2n} |P_t a(z) - P_t a(x)| dz.$$

Because $|P_t a(z) - P_t a(x)| \le \min(C, t^{-1}|x-z|)$, the right side of (2.45) is at most $Ct^{-n} \left(1 + k_n \left(\frac{|x-y|}{t}\right)\right)$, where $k_n(r) = |\log r|$ when $n = 1$, $k_n(r) = r^{-(n-1)}$ when $n > 1$. Therefore

(2.47)

$$|IIb(i)1| \le C \sum_j \int_0^\infty \iint \theta(x-y,t) t^{-n} \left(1 + k_n \left(\frac{|x-y|}{t}\right)\right) |\tilde{Q}_t^j f(y)||Q_t^j g(x)| dV_2$$

$$\le C \sum_j \int_0^\infty \int_{\mathbb{R}^n} \{\tilde{Q}_t^j f\}_{x,t} |Q_t^j g(x)| dx \frac{dt}{t}.$$

This is $\le C\|f\|_2 \|g\|_2$, by (2.8) and (2.10).

The argument for $IIb(i)2$ is very similar to what we did for S_1 in (2.25). We split $IIb(i)2$, which is given by

$$(2.48) \qquad -\sum_j \int_0^\infty \iint \theta(x-y,t) K_{P_t\,a(x)}(x,y)\tilde{Q}_t^j f(y) Q_t^j g(x) dV_2,$$

into two terms, corresponding to $\tilde{Q}_t^j f(y) = \tilde{Q}_t^j f(x) + (\tilde{Q}_t f(y) - \tilde{Q}_t^j f(x))$. The first is bounded by

$$C \int_0^\infty \int_{\mathbf{R}^n} |\tilde{Q}_t^j f(x)||Q_t^j g(x)| \frac{dxdt}{t}$$

[because $|\int \theta(x-y,t) K_{P_t a}(x,y) dy| \le C$, by (1.15) and (1.9)], and hence it is $\le C\|f\|_2\|g\|_2$. The second term is less than

$$C\sum_j \int_0^\infty \iint \theta(x-y,t)|x-y|^{-n} \left(\frac{|x-y|}{t} \right) \{\tilde{Q}_t^j f\}_{x,t} |Q_t^j g(x)| dV_2,$$

which is itself less than

$$C\sum \int_0^\infty \int \{\tilde{Q}_t^j f\}_{x,t} |Q_t^j g(x)| dV_2 \le C\|f\|_2\|g\|_2.$$

That takes care of $IIb(i)$. The computations for $IIa(i)$ are similar, and we omit them.

This leaves $IIa(ii)$ and $IIb(ii)$, which are the far-away parts of IIa and IIb. These are treated in essentially the same way that U_j and V_j in (2.30) and (2.31) were. First $\tilde{Q}_t^j f(y)$ is rewritten as $t\frac{\partial}{\partial y_j} \tilde{Q}_t f(y)$, and then the $\frac{\partial}{\partial y_j}$ is integrated by parts. The $\frac{\partial}{\partial y_j}$ then hits either the θ or the K, and in both cases there is adequate decay to prove that

$$|IIa(ii)| + |IIb(ii)| \le C\|f\|_2\|g\|_2.$$

The details are so similar to what we did before that we omit them.

References

[CDM] R.R. Coifman, G. David, and Y. Meyer, *La solution des conjectures de Calderón*, Adv. in Math. 48(1983), 144-148.

[CM] R.R. Coifman and Y. Meyer, <u>Au-delà des opérateurs pseudo-differentiels</u>, Astérisque 57, Société Mathématique de France, Paris, 1978.

[JL] J.L. Journé, <u>Calderón-Zygmund Operators, Pseudodifferential Operators, and the Cauchy Integral of Calderón</u>, Lecture Notes in Math., Springer-Verlag **999**, 1983.

Dept. of Mathematics
Yale University
New Haven, CT 06520

Dept. of Mathematics
Rice University
Houston, TX 77251

ON OPTIMAL RECOVERY OF MULTIVARIATE
PERIODIC FUNCTIONS

Dinh Dung
Institute of Computer Science
Lieu Giai, Ba Dinh, Hanoi, Vietnam

1. INTRODUCTION

Let X be a normed linear space of functions defined on the torus $T^d := [-\pi,\pi]^d$ and $W \subset X$. For a collection of points $\{x^1,\ldots,x^k\} \subset T^d$ and a mapping $P_k(t_1,\ldots,t_k)$ from R^k into a linear manifold in X of dimensions at most k, one can naturally consider recovering $f \in W$ from its values $f(x^1),\ldots,f(x^k)$ by the element $P_k(f(x^1),\ldots,f(x^k))$. We introduce the optimal recovery

$$R_n(W,X) := \inf_{\substack{f \in W}} \sup \|f - P_k(f(x^1),\ldots,f(x^k))\|_X ,$$

where the lower bound is taken over all collections $\{x^1,\ldots,x^k\}$ and P_k with $k \le n$.

In this paper we shall discuss the asymptotic degree of $R_n(SH_p^r,L_q)$ for $r \in R_+^d$ and $1 \le p < q \le \infty$, where SH_p^r is the unit ball in the space H_p^r of functions satisfying the mixed Hölder condition r in L_p (see a definition in Section 2) and $L_t := L_t(T^d)$, $1 \le t \le \infty$. Lattices and methods were constructed for recovering multivariate functions with a bounded mixed derivative from their values at these lattices [S], [HW], [T2] . In particular, in [T2] the error in recovering functions of classes SH_p^r coincides in power scale with the degree of the corresponding n-widths. This paper is also closely related to the problems of n-widths and of best approximations by trigonometric polynomials of so-called hyperbolic crosses (cf. [D1],[T1]).

The main result of this paper reads as follows. Let $r = (r_1,\ldots,r_d)$ and $0 < r_1 = \ldots = r_\nu = r_{\nu+1} < r_{\nu+2} \le \ldots \le r_d$ $(0 \le \nu \le d-1)$. Then, for $1 \le p < q \le \infty$:

$$(1) \qquad R_n(SH_p^r,L_q) \ll n^{-r_1+1/p-1/q}(\log n)^{\nu(r_1-1/p+2/q)} ,$$

and for $1 < p < q \leq 2$:

(2) $\qquad R_n(SH_p^r, L_q) \asymp n^{-r_1 + 1/p - 1/q}(\log n)^{\nu(r_1 - 1/p + 2/q)}$.

Here and below we use $\ll$ and $\asymp$ for notation of inequality and equivalence between asymptotic degrees (orders) (cf., e.g., [D1]) .

We conclude the Introduction with a short outline of the following sections. In Section 2 we introduce Hölder spaces of a mixed smoothness. A description of these spaces by harmonic decomposition and the corresponding equivalence of seminorms are proved. As preliminaries, we also establish some properties of trigonometric polynomials and of the sum convolution operator S_m . In Section 3 we prove (1-2). For this purpose we construct a lattice and a linear method for recovering functions of SH_p^r from their values at this lattice, using the above mentioned harmonic decomposition and equivalence of seminorms.

2. HOLDER SPACES AND HARMONIC DECOMPOSITION

First we introduce Hölder spaces of a mixed smoothness. For natural l and $h \in T^d$ the finite difference operator Δ_h^l is defined by

$$\Delta_h^l := \Delta_h^{l-1} \circ \Delta_h \; , \qquad \Delta_h := \prod_{i=1}^{d} \Delta_{h_i} \quad ,$$

$$(\Delta_{h_i} f)(x) := f(x_1, \ldots, x_i + h_i, \ldots, x_d) - f(x) \quad ,$$

where x_i denotes the i-th coordinate of x . If $1 \leq p \leq \infty$, $r \in R_+^d$, the Hölder space H_p^r [T1] consists of all functions f on T^d with zero mean value in each variable such that the seminorm

$$\| f \|_{H_p^r} := \sup_h \; \prod_{i=1}^{d} |h_i|^{-r_i} \| \Delta_h^l f \|_p$$

is finite for some $l > \max \{r_i : 1 \leq i \leq d\}$ where $R_+^d := \{x \in R^d : x > 0\}$ and the inequality $x > y$ $(x \geq y)$ for $x, y \in R^d$ is understood as $x_i > y_i$ $(x_i \geq y_i)$, $i = 1, \ldots, d$. The definition does not depend on the choice of values of l (see (11)). Functions in the space H_p^r have, in some sense, common smoothness r . Throughout this paper r will be fixed and without loss of generality we shall assume that for some $0 \leq \nu \leq d-1$

$$0 < r_1 = \cdots = r_\nu = r_{\nu+1} < r_{\nu+2} \leq \cdots \leq r_d .$$

In order to formulate and prove a harmonic decomposition and an equivalence of seminorms for H_p^r we introduce the sum convolution operator S_m and establish some properties of S_m and of trigonometric polynomials.

Let

$$F_m(x) := \prod_{i=1}^d \varphi_{m_i}(x_i) , \quad m \in N^d ,$$

be de la Vallée Pussin's kernel of d variables where

$$\varphi_k(t) := 1 + 2 \sum_{j=1}^k \cos jt + 2 \sum_{j=k+1}^{2k} \frac{2k-j}{j} \cos jt .$$

The sum convolution operator S_m is defined by

$$S_m f := \prod_{i=1}^d (3m_i)^{-1} \sum_{k \in D_{3m}} f(hk) F_m(.-hk) , \quad h = 2\pi/3m ,$$

for functions f on T^d where $D_s := \{k \in Z^d : 0 \leq k < s\}$ for $s \in N^d$; $1/x := (1/x_1, \ldots, 1/x_d)$, $xy := (x_1 y_1, \ldots, x_d y_d)$ for $x, y \in R^d$.

We shall need the following properties:

(3) $\|F_m\|_1 \ll 1 , \quad m \in N^d$

(4) $(S_m f)(hk) = f(hk) , \quad 0 \leq k < 3m \quad (h = 2\pi/3m)$

(5) $S_m f = f$ for any $f \in T_m ,$

where T_m denotes the set of trigonometric polynomials of order at most m_i at the variable x_i, $i = 1, \ldots, d$. The relations (3-4) can be easily verified. The latter one follows from a more generalized assertion. Namely, if $m, n, s \in N^d$ and $m+n < s$, then the equality

(6) $$f * g = \prod_{i=1}^d s_i^{-1} \sum_{k \in D_s} f(hk) g(.-hk), \quad h = 2\pi/s .$$

holds for any $f \in T_m$ and $g \in T_n$. Indeed, in view of the formular

$$f * g = \sum_k f_k g_k e^{i\langle k, \cdot \rangle}$$

where $\langle x, y \rangle = x_1 y_1 + \ldots + x_d y_d$, f_k and g_k are the k-th Fourier coefficients of f and g, it is sufficient to prove that for any $g \in T_n$ and $f = e^{i\langle k, \cdot \rangle}$, $k \leq m$, the right side of (6) is equal to $g_k e^{i\langle k, \cdot \rangle}$. This fact can be checked directly by replacing $g(.-hk)$ with its Fourier series. Since $F_m * f = f$ for any $f \in T_m$ the above proved assertion implies (5).

For a sequence $\{a_k\}_{k \in D_s}$ of real numbers we introduce the norm

$$\|\{a_k\}\|_{p,s}^p := (2\pi)^d \prod_{i=1}^d s_i^{-1} \sum_{k \in D_s} |a_k|^p , \quad 1 \leq p < \infty ,$$

the sum norm changed to the max norm when $p = \infty$.

For any $f \in T_m$ we have

(7) $$\sup_x \|\{f(x-hk)\}\|_{p,s} \leq \prod_{i=1}^d (1+h_i m_i)^{1/p} \|f\|_p , \quad h = 2\pi/s .$$

The case $p = \infty$ of this inequality is obvious. The case $p = 1$ can be proved in a way similar to the proof of Theorem 3.3.2 [N] establishing an analogous inequality for functions of exponential type. The case $1 < p < \infty$ follows from the cases $p = 1, \infty$ by interpolation properties of spaces L_p.

Lemma 1. If $1 \leq p \leq \infty$, then for any $m \in N^d$ and $f \in C(T^d)$

$$\|S_m f\|_p \ll \|\{f(hk)\}\|_{p,3m} , \quad h = 2\pi/3m .$$

Proof. We prove the lemma for $1 \leq p < \infty$, the case $p = \infty$ can be proved in a similar way. By the Hölder inequality we have

$$\prod_{i=1}^d (3m_i)^p |(S_m f)(x)|^p$$

$$\leq \left(\sum_{k \in D_{3m}} |F_m(x-hk)| \right)^{p/p'} \sum_{k \in D_{3m}} |f(hk)|^p |F_m(x-hk)| ,$$

where $1/p + 1/p' = 1$. Hence by (3) and (7) we get

$$\|S_m f\|_p^p \leq (\sup_x \|\{F_m(x-hk)\}\|_{1,3m})^{p/p'} \|F_m\|_1 \|\{f(hk)\}\|_{p,3m}^p$$

$$\ll \|\{f(hk)\}\|_{p,3m}^p .$$

The proof is complete.

Corollary 1. For any $f \in C(T^d)$ we have

$$\lim_{\substack{m_i \to \infty \\ i=1,\ldots,d}} \|f - S_m f\|_{C(T^d)} = 0 .$$

Proof. Lemma 1 yields

(8)
$$\|S_m\|_{C(T^d) \to C(T^d)} \leq c$$

for some positive constant c . Given an arbitrary $\varepsilon > 0$, we find $g \in T_{m^\circ}$ for some $m^\circ \in N^d$ such that $\|f - g\|_{C(T^d)} < \varepsilon/(1+c)$. By (5) and (8) we have for any $m \geq m^\circ$

$$\|f - S_m f\|_{C(T^d)} \leq \|f - g\|_{C(T^d)} + \|S_m(f - g)\|_{C(T^d)}$$

$$\leq (1+c)\|f - g\|_{C(T^d)} < \varepsilon .$$

This proves the corollary.

From Lemma 1 and (5), (7) we obtain the following

Corollary 2. If $1 \leq p \leq \infty$, then for any $m \in N^d$ and $f \in T_m$

$$\|f\|_p \ll \|\{f(hk)\}\|_{p,3m} , \quad h = 2\pi/3m .$$

Corollary 3. If $1 \leq p \leq \infty$, then for any $f \in T_n$, $n \geq m$, $n,m \in N^d$

$$\|S_m f\|_p \ll \prod_{i=1}^{d} (n_i/m_i)^{1/p} \|f\|_p .$$

Corollary 2 is a modification of Marcinkiewicz' theorem (cf.[Z]).

We next describe two harmonic decompositions of spaces H_p^r. Let U_k and V_k be the unidimensional operators defined by

$$U_1 := S_1 , \quad U_k := S_{2^{k-1}} - S_{2^{k-2}} , \quad k = 2,3,\ldots$$

$$V_1 := I_1 , \quad V_k := I_{2^{k-1}} - I_{2^{k-2}} , \quad k = 2,3,\ldots$$

where $I_m f := F_m * f$ for $m \in N^d$. The mixed multidimensional operators U_k and V_k for $k \in N^d$ are defined by

$$U_k := \prod_{i=1}^{d} U_{k_i} , \quad V_k := \prod_{i=1}^{d} V_{k_i}$$

where U_{k_i} and V_{k_i} are the unidimensional operators at the variable x_i.

Note that for any $m \in N^d$

$$S_{2^m} = \sum_{k \le m} U_k ; \quad I_{2^m} = \sum_{k \le m} V_k$$

where $2^x := (2^{x_1}, \ldots, 2^{x_d})$ for $x \in R^d$. From the first equality and Corollary 1 it follows that every $f \in C(T^d)$ can be represented by the series

$$(9) \qquad f(x) = \sum_k (U_k f)(x) , \quad x \in T^d ,$$

converging uniformly on T^d. From the second one and

$$\lim_{\substack{m_i \to \infty \\ i = 1,\ldots,d}} \| f - I_m f \|_p = 0 , \quad f \in L_p \quad (1 \le p \le \infty)$$

it follows that every $f \in L_p$ acn be represented by the L_p-converging series

$$(10) \qquad f = \sum_k V_k f .$$

For every $f \in H_p^r$ the following equivalence of seminorms holds [T1]

$$(11) \qquad \| f \|_{H_p^r} \asymp \sup_k 2^{\langle r,k \rangle} \| V_k f \|_p .$$

Lemma 2. If $1 \le p \le \infty$ and $r_1 > 1/p$, then the following equivalence of seminorms holds for $f \in H_p^r$

$$\|f\|_{H_p^r} \asymp \sup_k 2^{\langle r,k \rangle} \|U_k f\|_p .$$

Proof. From the condition $r_1 > 1/p$ it follows that H_p^r is compactly embedded into $C(T^d)$. Thus, by (9) we have for each $f \in H_p^r$

$$(U_k f)(x) = \sum_s (U_k V_s f)(x) .$$

Using (5), one can see that $U_k V_s f \equiv 0$ whenever the inequality $s > k - e$ does not hold. Here and below $e = (1,1,\ldots,1)$. Therefore,

$$(12) \qquad \|U_k f\|_p \le \sum_{s > k-e} \|U_k V_s f\|_p .$$

Corollary 3 implies for $s \ge k$

$$\|U_k V_s f\|_p \ll 2^{\langle e/p, s-k \rangle} \|V_s f\|_p .$$

Hence by (12) and (11) we obtain

$$2^{\langle r,k \rangle} \|U_k f\|_p \ll 2^{\langle r,k \rangle} \sum_{s > k-e} 2^{\langle e/p, s-k \rangle} 2^{-\langle r,s \rangle} \|f\|_{H_p^r}$$

$$(13) \qquad \ll \|f\|_{H_p^r} 2^{\langle r-e/p, k \rangle} \sum_{s > k-e} 2^{-\langle r-e/p, s \rangle} .$$

In virtue of the inequality $r_1 > 1/p$ the series in the latter expression does not exceed a multiple of $2^{\langle r-e/p, k \rangle}$. By use of this estimate we get from (13)

$$(14) \qquad \sup_k 2^{\langle r,k \rangle} \|U_k f\|_p \ll \|f\|_{H_p^r}$$

The inverse asymptotic inequality can be proved in the same way by replacing roles of U_k and V_k by one other. The proof is complete.

An inequality similar to (14) was proved [T2] for a harmonic decomposition with a different method.

Finally, we shall need an inequality between L_p-norm and L_q-norm

Let f be a function on T^d, represented by the series

$$f(x) = \sum_{k \in N^d} f_k(x) , \quad f_k \in T_{2^k} ,$$

satisfying the condition

$$\sum_{k \in N^d} (2^{(1/p-1/q)|k|} \|f_k\|_p)^q \quad , \quad 1 \le p < q < \infty .$$

Then $f \in L_q$ and

(15) $$\|f\|_q^q \le C \sum_{k \in N^d} (2^{(1/p-1/q)|k|} \|f_k\|_p)^q$$

with some positive constant C independent of f. Here $|k| := k_1 +$ ... $+ k_d$ for $k \in N^d$. This inequality was proved in [T1] for a harmonic decomposition and in [D2] in the general case.

3. OPTIMAL RECOVERY

To estimate $R_n(SH_p^r, L_q)$ we shall construct a linear method for recovering functions of SH_p^r on basic of the harmonic decomposition (9) where

$$SH_p^r := \left\{ f \in H_p^r : \|f\|_{H_p^r} \le 1 \right\}$$

the unit ball of H_p^r. We define the recovery operator Q_s for natural s by

$$Q_s f := \sum_{k \in G_s} U_k f$$

for functions f on T^d, where $G_s := \{k \in N^d : \langle r,k \rangle \le s\}$. A similar operator was constructed in [T2]. Note that $Q_s f$ is a trigonometric polynomial and completely constructed from values of f at the lattice

$$L_s := \left\{ x_{kj} = j2^{-k}/3 : k \in G_s , \quad j \in D_{3 \cdot 2^k} \right\} .$$

One can easily compute the following asymptotic degree of the number of points in L_s :

(16) $$\text{card } L_s \asymp 2^{s/r_1} s , \quad s \to \infty .$$

By the harmonic decomposition (9) and inequality (15) for $f \in C(T^d)$ the error in recovering f by $Q_s f$ from its values L_s can be estimated as follows:

$$(17) \qquad \|f - Q_s f\|_q^q \ll \sum_{k \in N^d \setminus Q_s} (2^{(1/p-1/q)|k|} \|U_k f\|_p)^q$$

for $1 \leq p < q < \infty$.

For a given natural n let $m := \max \{s : \text{card } L_s \leq n\}$.

Theorem 1. If $1 \leq p < q < \infty$ and $r_1 > 1/p$, then

$$R_n(SH_p^r, L_q) \ll \sup_{f \in SH_p^r} \|f - Q_m f\|_q$$

$$\ll n^{-r_1 + 1/p - 1/q} (\log n)^{\nu(r_1 - 1/p + 1/q)} .$$

Proof. Lemma 2 and (17) give for each $f \in SH_p^r$

$$(18) \qquad \|f - Q_m f\|_q^q \ll \sum_{k \in N^d \setminus G_m} 2^{-q\langle r - e/p + e/q, k \rangle} =: A_m .$$

By a simple computation we get

$$(19) \qquad A_m \asymp 2^{-am} m \quad , \quad a = q(r_1 - 1/p + 1/q)/r_1 .$$

On the other hand, by the difinition of m and (16) we have $2^{m/r_1} m \asymp n$, $n \to \infty$. Hence by (18-19) we obtain

$$\|f - Q_m f\|_q^q \ll A_m \asymp n^{-ar_1} (\log^\nu m)^{ar_1 + 1} .$$

This completes the proof.

Theorem 2. Under the hypotheses of Theorem 1 if additionally $1 < p < q \leq 2$, then

$$(20) \qquad R_n(SH_p^r, L_q) \asymp n^{-r_1 + 1/p - 1/q} (\log n)^{\nu(r_1 - 1/p + 2/q)} .$$

Proof. The upper bound of (20) is contained in Theorem 1. We now establish the lower bound. From the definition of $R_n(W, X)$ we get

$$R_n(W,X) \geq d_n(W,X)$$

where $d_n(W,X)$ denotes the n-width of W in X (cf., e.g., [D1] for the definition of n-width). Thus, using the estimate

$$d_n(SH_p^r,Lq) \gg n^{-r_1+1/p-1/q}(\log n)^{\nu(r_1-1/p+2/q)}$$

proved in [G] where a different equivalent definition of SH_p^r was used, we obtain the lower bound of (20). The theorem is proved.

REFERENCES

[D1] Dinh Dung, Approximation by trigonometric polynomials of functions of several variables on the torus, Mat. Sb. 131(1986),251-271.

[D2] Dinh Dung, Approximation of smooth functions of several variables by means of harmonic analysis (Doctor's Dissertation), Moscow 1985.

[G]   E.M. Galeev, Kolmogorov's widths of some classes of periodic functions of one or several variables, Izv. Akad. Nauk SSSR Ser. Mat. 54(1990), 418-430.

[HW] Hua Lookeng and Wang Yuan, Applications of Number Theory to Numerical Analysis, Springer, Berlin 1981.

[N]   S.M. Nikol'skii, Approximation of Functions of Several Variables and Embedding Theorems, Springer,Berlin 1975.

[S]   S.A. Smolyak, Quadrature anf interpolation formular for tensor products of certain classes of functions, Dokl. Akad. Nauk SSSR 148(1963), 1042-1045.

[T1] V.N. Temlyakov, Approximation of functions with a bounded mixed derivative, Trudy Mat. Inst. Steklov 178(1986), 3-112.

[T2] V.N. Temlyakov, Approximate recovery of periodic functions of several variables, Mat. Sb. 128(1985), 256-268.

[Z]   A. Zygmund, Trigonometric Series II, Cambridge Univ. Press 1959.

A CLASS OF WEIGHTED INEQUALITIES

Salah A.A. Emara*

The American University in Cairo

ABSTRACT. Given indices p and q, $1 \leq p, q \leq \infty$, T a quasilinear operator bounded between weighted interpolation A_{w_i,q_i} to $B_{\bar{w}_i,\bar{q}_i}$-spaces, $i = 0, 1$, $1 \leq q_i, \bar{q}_i \leq \infty$, where w, $\bar{w}$ are weight functions belonging to some class of functions B_K. We give conditions on pairs of functions u and v which are sufficient that

$$\left(\int_0^\infty [u(t)K(t, Tf; B)]^q \, dt \right)^{1/q} \leq C \left(\int_0^\infty [v(t)K(t, f; A)]^p \, dt \right)^{1/p},$$

holds, where C is a constant independent of f and $K(.,.;.)$ is the Peetre K-functional.

The following result is the well-known weighted Hardy inequality and the dual inequality.

Lemma 1 ([2],[1],[8]). *Suppose* u, v *and* f *are non-negative functions on* $\mathbf{R}^+$ *and* $1 \leq p \leq q \leq \infty$. *Then*

$$(1) \qquad \left[\int_0^\infty \left[u(t) \int_0^t f(y) \, dy \right]^q dt \right]^{1/q} \leq C \left[\int_0^\infty [v(t)f(t)]^p \, dt \right]^{1/p},$$

if and only if

$$(2) \qquad \sup_{s>0} \left(\int_s^\infty u(t)^q dt \right)^{1/q} \left(\int_0^s v(t)^{-p'} dt \right)^{1/p'} \equiv A < \infty.$$

Moreover, $A \leq C \leq p^{1/q}(p')^{1/p'} A$ *if* $p \neq 1$, *where* C *is the least constant for which* (1) *holds; if* $p = 1$, *then* $A = C$. *Note that in this case the second integral of* (2) *is interpreted to be the essential supremum of* $1/v$ *for* $t < s$. *(A similar interpretation is made for the first integral of* (2) *if* $q = \infty$).

The dual inequality

$$\left[\int_0^\infty \left[u(t) \int_t^\infty f(y) \, dy \right]^q dt \right]^{1/q} \leq C \left[\int_0^\infty [v(t)f(t)]^p \, dt \right]^{1/p}$$

holds if and only if

$$\sup_{s>0} \left(\int_0^s u(t)^q dt \right)^{1/q} \left(\int_s^\infty v(t)^{-p'} dt \right)^{1/p'} \equiv B < \infty.$$

Mazja [8] and independently Sawyer [9] extended the above result to the range $1 < q < p < \infty$, in fact Sawyer extended the result to the range $0 < q < p$, $p > 1$

---

*Research supported in part by The American University in Cairo.

with different but equivalent weight conditions. We state here their result in the form of the weight conditions of Mazja.

Lemma 2 ([8],[6],[9]).
(a) Suppose u, v and f are non-negative functions on $\mathbf{R}^+$ and $1 \leq q < p < \infty$. Then

(3)
$$\left[\int_0^\infty \left[u(t) \int_0^t f(y)\,dy \right]^q dt \right]^{1/q} \leq C \left[\int_0^\infty [v(t)f(t)]^p\,dt \right]^{1/p},$$

if and only if

(4)
$$\left[\int_0^\infty \left[\left(\int_x^\infty u(y)^q dy \right)^{1/q} \left(\int_0^x v(y)^{-p'} dy \right)^{1/q'} \right]^r v(x)^{-p'} dx \right]^{1/r} < \infty,$$

where $1/r = 1/q - 1/p$.
In case $q = 1 < p$, condition (4) takes the form

$$\left[\int_0^\infty \left(\int_x^\infty u(y)\,dy \right)^{p'} v(x)^{-p'} dx \right]^{1/p'} < \infty.$$

(b) If u, v, f and r are as above, then

(5)
$$\left[\int_0^\infty \left[u(t) \int_t^\infty f(y)\,dy \right]^q dt \right]^{1/q} \leq C \left[\int_0^\infty [v(t)f(t)]^p\,dt \right]^{1/p}$$

if and only if

(6)
$$\left[\int_0^\infty \left[\left(\int_0^x u(y)^q dy \right)^{1/q} \left(\int_x^\infty v(y)^{-p'} dy \right)^{1/q'} \right]^r v(x)^{-p'} dx \right]^{1/r} < \infty.$$

If $q = 1 < p$, condition (6) is modified in a similar way as in (a).

Note that in the limiting case $q = p$ the integral in (4) takes the form (2) and similarly the integral (6) takes the corresponding form of the weight conditions of the dual operator in Lemma 1.

Lemma 3 ([3]). Let $0 < s < 1$ and f a non-negative and g a positive continuous function.
(i) If f is increasing on $[0, \infty)$ and g decreasing on $[0, \infty)$, such that $\lim_{t \to \infty} g(t) = 0$, then for $t \geq 0$

(7)
$$s \int_t^\infty f(x)\,d(-g(x)) \leq \left[\int_t^\infty f(x)^s d(-g(x)^s) \right]^{1/s}$$

holds.
(ii) If f is decreasing on $[0, \infty)$ and g increasing on $[0, \infty)$, such that $\lim_{t \to 0} g(t) = 0$, then for $t \leq \infty$

(8)
$$s \int_0^t f(x)\,d(g(x)) \leq \left[\int_0^t f(x)^s d(g(x)^s) \right]^{1/s}$$

holds.

Part (i) follows from Calderón and Scott [3, Lemma 6.1] and Part (ii) follows at once from the first on substituting x by $1/x$. We give the proof of Part (i) here only for completeness.

Proof.

(i) Since the case $t = 0$ is known [3], we assume $t > 0$. Moreover, it suffices to prove the result when the integral on the right of (7) is finite. Let $I_{s,t}$ denote the right side of (7) then for $y \geq t > 0$

$$I_{s,t} \geq \left[\int_y^\infty f(x)^s d(-g(x)^s) \right]^{1/s} \geq f(y)g(y),$$

so that $f(y) \leq I_{s,t}/g(y)$. Using this estimate and integrating we get

$$\int_t^\infty f(x)\, d(-g(x)) = \int_t^\infty f(x)^s f(x)^{1-s} d(-g(x))$$

$$\leq \int_t^\infty f(x)^s I_{s,t}^{1-s} g(x)^{s-1} d(-g(x))$$

$$= s^{-1} I_{s,t}^{1-s} \int_t^\infty f(x)^s\, d(-g(x)^s) = s^{-1} I_{s,t},$$

which proves the first part of the lemma.

We now define the weight class B_K introduced by Kalugina [7].

Definition 4 ([7]). *A continuous non-decreasing function $w : \mathbf{R}^+ \to \mathbf{R}^+$ belongs to the B_K, if*

$$\int_0^\infty \min(1, 1/t)\tilde{w}(t)t^{-1}dt < \infty,$$

where $\tilde{w}(s) = \sup_{y>0} w(sy)/w(y)$ and $\tilde{w}(s) < \infty$ for $s > 0$.

Proposition 5 [4]. *If $w \in B_K$ then*
(i) $\underset{\sim}{w}(s)\tilde{w}(1/s) = 1$, where $\underset{\sim}{w}(s) = \inf_{t>0} w(st)/w(t)$.

(ii) $0 < \underset{\sim}{w}(s)w(t) \leq w(st) \leq \tilde{w}(s)w(t)$.

(iii) $\tilde{w}$ and $\underset{\sim}{w}$ are non-decreasing and $\tilde{w}(1) = \underset{\sim}{w}(1) = 1$.

(iv) For any $p > 0$

$$\left[\int_0^\infty [\min(1, 1/t)\tilde{w}(t)]^p\, t^{-1}dt \right]^{1/p} < \infty,$$

with the usual modification if $p = \infty$.
(v) There exist constants $A, B > 0$ such that for all $s > 0$,

$$A \leq s^{-1}w(s) \left[\int_0^s [t/w(t)]^p\, t^{-1}dt \right]^{1/p} \leq B, \quad p > 0.$$

In fact $A = p^{-1/p}$ *and* $B = [\int_0^\infty [\tilde{w}(t)/t]^p \, t^{-1} dt]^{1/p}$, *if* $p < \infty$.

(vi) *There are positive constants* C, D *such that*

$$C \le w(s) \left[\int_s^\infty [1/w(t)]^p \, t^{-1} dt \right]^{1/p} \le D, \quad p > 0.$$

Here, $C = [\int_1^\infty [1/\tilde{w}(t)]^p \, t^{-1} dt]^{1/p}$ *and* $D = [\int_0^1 \tilde{w}(t)^p t^{-1} dt]^{1/p}$, *if* $p < \infty$.

Clearly, if $0 < \theta < 1$ then $w(t) = t^\theta \in B_K$. Also, as Gustavsson [4] has shown, the function $t^\beta / \log(1 + t^\alpha) \in B_K$ if $0 < \alpha < \beta < 1$.

The weight class B_ψ is now defined as follows:

Definition 6 ([7]). *B_ψ consists of all non-negative continuously differentiable functions w on $\mathbf{R}^+$ such that*

$$\sup_{t>0} \, tw'(t)/w(t) = \beta < 1 \quad \text{and} \quad \inf_{t>0} \, tw'(t)/w(t) = \alpha > 0.$$

It is not difficult to see ([4, Prop.1.2]) that $B_\psi \subset B_K$.

Again $w(t) = t^\theta, 0 < \theta < 1$ is in B_ψ and also $w(t) = t^\alpha(\log(1 + t^\gamma))$, $0 < \alpha < 1, \theta$ real and γ in a sufficiently small neighbourhood of zero (Gustavsson [4]).

Definition 7 ([7],[4]). *Let $w \in B_\psi$ and (A_0, A_1) an interpolation couple. The weighted intermediate spaces $A_{w,p} = (A_0, A_1)_{w,p}, 0 < p \le \infty$ with weight w consists of all $f \in A_0 + A_1$ such that the functional*

$$\|f\|_{A_{w,p}} = \begin{cases} \left(\int_0^\infty [K(t, f; A_0, A_1)/w(t)]^p t^{-1} dt \right)^{1/p}, & 0 < p < \infty \\ \sup_{t>0} K(t, f; A_0, A_1)/w(t), & p = \infty \end{cases}$$

is finite.

Remark 8. If $w_i \in B_\psi, i = 0, 1$ and $\tau(t) = w_1(t)/w_0(t)$ satisfies $t\tau'(t)/\tau(t) \ge \alpha > 0$, for all $t > 0$, then τ has an inverse and $\lim_{t \to 0} \tau(t) = 0$, $\lim_{t \to \infty} \tau(t) = \infty$.

Clearly the above inequality implies $\tau'(t) > 0$ so τ has an inverse. Also if $0 < t < 1$ then the inequality implies

$$\log \tau(1) - \log \tau(t) = \int_t^1 \tau'(s)/\tau(s) \, ds \ge \alpha \int_t^1 s^{-1} ds = -\alpha \log t$$

which implies $\tau(t) \le \tau(1)t^\alpha$ so that $\tau(t) \to 0$ as $t \to 0$. Similarly if $1 < t < \infty$ then

$$\log(\tau(t)/\tau(1)) = \int_1^t \tau'(s)/\tau(s) \, ds \ge \alpha \log t$$

which implies $\tau(t) \to \infty$ as $t \to \infty$.

Now, we state and prove the weighted inequality for the K-functional for $1 \le p \le q \le \infty$.

Theorem 9. Suppose $A = (A_0, A_1)$, $B = (B_0, B_1)$ are interpolation couples, $w_i, \bar{w}_i \in B_\psi$, $i = 0, 1$ and $\tau = w_1/w_0, \bar{\tau} = \bar{w}_1/\bar{w}_0$ satisfy $|t\tau'(t)/\tau(t)| \ge \alpha > 0$ and $t\bar{\tau}'(t)/\bar{\tau}(t) \ge \bar{\alpha} > 0$. Further, let $\eta = \tau^{-1}$, $\bar{\eta} = \bar{\tau}^{-1}, \sigma = \bar{\eta} \circ \tau$ and T be a quasi-linear operator satisfying

$$T : A_{w_i, q_i} \to B_{\bar{w}_i, \bar{q}_i},$$

$i = 0, 1$, $1 \le q_i, \bar{q}_i \le \infty$.

In case $t\tau'(t)/\tau(t) \ge \alpha > 0$ and the weights u, v satisfy

(9) $$\sup_{s>0} \left[\int_{\sigma(s)}^\infty [u(t)\bar{w}_0(t)]^q \, dt \right]^{1/q} \left[\int_0^s [v(t)tw_0(t)]^{-p'} \, dt \right]^{1/p'} < \infty$$

and

(10) $$\sup_{s>0} \left[\int_0^{\sigma(s)} [u(t)\bar{w}_1(t)]^q \, dt \right]^{1/q} \left[\int_s^\infty [v(t)tw_1(t)]^{-p'} \, dt \right]^{1/p'} < \infty$$

with $1 \le p \le q \le \infty$, then there is a $C > 0$ such that

(11) $$\left(\int_0^\infty [u(t)K(t, Tf; B)]^q \, dt \right)^{1/q} \le C \left(\int_0^\infty [v(t)K(t, f; A)]^p \, dt \right)^{1/p}.$$

If $-t\tau'(t)/\tau(t) \ge \alpha > 0$, (11) holds provided (9) and (10) are satisfied with the ranges of integrations of the first two integrals in (9) and (10) are interchanged.

Proof. From Theorem 2.1 of Heinig [5] it follows that:
If $X_i = A_{w_i, q_i}, i = 0, 1; (X_0, X_1) = D$, then

(12) $$K(t, f; D) \sim \left\{ \int_0^{\eta(t)} [K(s, f; A)/w_0(s)]^{q_0} \, s^{-1} ds \right\}^{1/q_0}$$

$$+ t \left\{ \int_{\eta(t)}^\infty [K(s, f; A)/w_1(s)]^{q_1} \, s^{-1} ds \right\}^{1/q_1}.$$

Writing $Y_i = B_{\bar{w}_i, \bar{q}_i}, i = 0, 1$ and $(Y_0, Y_1) = E$, this shows that

(13) $$K(t, f; E) \sim \left\{ \int_0^{\bar{\eta}(t)} [K(s, f; B)/\bar{w}_0(s)]^{\bar{q}_0} \, s^{-1} ds \right\}^{1/\bar{q}_0}$$

$$+ t \left\{ \int_{\bar{\eta}(t)}^\infty [K(s, f; B)/\bar{w}_1(s)]^{\bar{q}_1} \, s^{-1} ds \right\}^{1/\bar{q}_1}.$$

Since $T : X_i \to Y_i$ is bounded, there exist numbers $M_i > 0$ such that, $\|Tf_i\|_{Y_i} \le M_i \|f_i\|_{X_i}, f_i \in X_i, i = 0, 1$ and therefore we have $K(t, Tf; E) \le M_0 K(tM_1/M_0, f; D)$. We may take without loss of generality $M_1/M_0 = 1$, so that

(14) $$K(t, Tf; E) \le CK(t, f; D).$$

Following the argument of [5, Theorem 2.1] we now prove the estimates

$$K(\bar{\eta}(t), Tf; B)/\bar{w}_0(\bar{\eta}(t)) \leq C \left[\int_0^{\bar{\eta}(t)} [K(s, Tf; B)/\bar{w}_0(s)]^{\bar{q}_0} s^{-1} ds \right]^{1/\bar{q}_0}$$

and

(15) $$K(\bar{\eta}(t), Tf; B)/\bar{w}_0(\bar{\eta}(t)) \leq Ct \left[\int_{\bar{\eta}(t)}^{\infty} [K(s, Tf; B)/\bar{w}_1(s)]^{\bar{q}_1} s^{-1} ds \right]^{1/\bar{q}_1}.$$

From Proposition 5(v) with p, s and $w(s)$ replaced by $\bar{q}_0, \bar{\eta}(s)$ and $\bar{w}_0(s)$, respectively, we get

$$A \leq \bar{w}_0(\bar{\eta}(s))/\bar{\eta}(s) \left\{ \int_0^{\bar{\eta}(s)} [t/\bar{w}_0(t)]^{\bar{q}_0} t^{-1} dt \right\}^{1/\bar{q}_0} \leq B,$$

that is

$$\int_0^{\bar{\eta}(s)} [t/\bar{w}_0(t)]^{\bar{q}_0} t^{-1} dt \geq C[\bar{\eta}(s)/\bar{w}_0(\bar{\eta}(s))]^{\bar{q}_0}.$$

Since $K(s, Tf)/s$ is decreasing then the previous inequality shows that

$$\int_0^{\bar{\eta}(t)} [K(s, Tf; B)/\bar{w}_0(s)]^{\bar{q}_0} s^{-1} ds \geq C[K(\bar{\eta}(t), Tf; B)/\bar{\eta}(t)]^{\bar{q}_0} \int_0^{\bar{\eta}(t)} [s/\bar{w}_0(s)]^{\bar{q}_0} s^{-1} ds$$

$$\geq C[K(\bar{\eta}(t), Tf; B)/\bar{w}_0(\bar{\eta}(t))]^{\bar{q}_0}$$

and therefore

(16) $$K(\bar{\eta}(t), Tf; B)/\bar{w}_0(\bar{\eta}(t)) \leq C \left\{ \int_0^{\bar{\eta}(t)} [K(s, Tf; B)/\bar{w}_0(s)]^{\bar{q}_0} s^{-1} ds \right\}^{1/\bar{q}_0}.$$

Similarly, from (vi) of Proposition 5 with p, s and $w(s)$ are replaced by $\bar{q}_1, \bar{\eta}(s)$ and $\bar{w}_1(s)$, respectively, we obtain

$$C \leq \bar{w}_1(\bar{\eta}(s)) \left\{ \int_{\bar{\eta}(s)}^{\infty} [1/\bar{w}_1(t)]^{\bar{q}_1} t^{-1} dt \right\}^{1/\bar{q}_1} \leq D,$$

that is

$$\int_{\bar{\eta}(s)}^{\infty} [1/\bar{w}_1(t)]^{\bar{q}_1} t^{-1} dt \geq C[1/\bar{w}_1(\bar{\eta}(s))]^{\bar{q}_1}.$$

Also since $K(t, Tf)$ is increasing we have

$$\int_{\bar{\eta}(t)}^{\infty} [K(s, Tf; B)/\bar{w}_1(s)]^{\bar{q}_1} s^{-1} ds$$

$$\geq K(\bar{\eta}(t), Tf; B)^{\bar{q}_1} \int_{\bar{\eta}(t)}^{\infty} [1/\bar{w}_1(s)]^{\bar{q}_1} s^{-1} ds$$

$$\geq C[K(\bar{\eta}(t), Tf; B)/\bar{w}_1(\bar{\eta}(t))]^{\bar{q}_1}$$

and hence we obtain

$$K(\bar{\eta}(t), Tf; B)/\bar{w}_1(\bar{\eta}(t)) \leq C \left\{ \int_{\bar{\eta}(t)}^{\infty} [K(s, Tf; B)/\bar{w}_1(s)]^{\bar{q}_1} s^{-1} ds \right\}^{1/\bar{q}_1}.$$

But $\bar{\tau}(t) = \bar{w}_1(t)/\bar{w}_0(t)$, ($t = \bar{\tau}(\bar{\eta}(t)) = \bar{w}_1(\bar{\eta}(t))/\bar{w}_0(\bar{\eta}(t))$), thus one gets (15). By combining (15) and (16) one obtains

$$K(\bar{\eta}(t), Tf; B)/\bar{w}_0(\bar{\eta}(t)) \leq C \left\{ \left[\int_0^{\bar{\eta}(t)} [K(s, Tf; B)/\bar{w}_0(s)]^{\bar{q}_0} s^{-1} ds \right]^{1/\bar{q}_0} \right.$$
$$\left. + t \left[\int_{\bar{\eta}(t)}^{\infty} [K(s, Tf; B)/\bar{w}_1(s)]^{\bar{q}_1} s^{-1} ds \right]^{1/\bar{q}_1} \right\}$$
$$\leq CK(t, Tf; E) \leq CK(t, f; D) \leq C \left\{ \left[\int_0^{\eta(t)} [K(s, f; A)/w_0(s)]^{q_0} s^{-1} ds \right]^{1/q_0} \right.$$
$$\left. + t \left[\int_{\eta(t)}^{\infty} [K(s, f; A)/w_1(s)]^{q_1} s^{-1} ds \right]^{1/q_1} \right\},$$

where the second, the third and the last inequalities follow from (13), (14) and (12), respectively. Hence

(17) $$K(\bar{\eta}(t), Tf; B)/\bar{w}_0(\bar{\eta}(t)) \leq C \left\{ \left[\int_0^{\eta(t)} [K(s, f; A)/w_0(s)]^{q_0} s^{-1} ds \right]^{1/q_0} \right.$$
$$\left. + t \left[\int_{\eta(t)}^{\infty} [K(s, f; A)/w_1(s)]^{q_1} s^{-1} ds \right]^{1/q_1} \right\}.$$

Lemma 3(ii) applied to the first integral of (17) and one gets

(18) $$\int_0^t [K(s, f; A)/w_0(s)]^{q_0} s^{-1} ds$$
$$= \int_0^t [K(s, f; A)/s]^{q_0} d \left(\int_0^s [x/w_0(x)]^{q_0} x^{-1} dx \right)$$
$$\leq C \left\{ \int_0^t s^{-1} K(s, f; A) d \left(\int_0^s [x/w_0(x)]^{q_0} x^{-1} dx \right)^{1/q_0} \right\}^{q_0}$$
$$= C \left\{ \int_0^t s^{-1} K(s, f; A) [s/w_0(s)]^{q_0} \left[\int_0^s [x/w_0(x)]^{q_0} x^{-1} dx \right]^{1/q_0 - 1} s^{-1} ds \right\}^{q_0}.$$

Next, from Proposition 5(v), it follows that

$$[s/w_0(s)]^{q_0} \leq C \int_0^s [t/w_0(t)]^{q_0} t^{-1} dt$$

and since $1/q_0 - 1 < 0$ we get

$$\left\{ \int_0^s [t/w_0(t)]^{q_0} t^{-1} dt \right\}^{1/q_0 - 1} \leq C[s/w_0(s)]^{q_0(1/q_0 - 1)}.$$

Substituting this estimate in the inner integral in the right side of (18) one sees that (18) yields

(19) $$\left\{ \int_0^t [K(s, f; A)/w_0(s)]^{q_0} s^{-1} ds \right\}^{1/q_0}$$
$$\leq C \int_0^t [K(s, f; A)/w_0(s)] s^{-1} ds, \quad 1 < q_0 < \infty.$$

Similarly, Lemma 3(i) applied to the second integral of (17) and we obtain

$$\int_t^\infty [K(s,f;A)/w_1(s)]^{q_1} s^{-1} ds = \int_t^\infty K(s,f;A)^{q_1} d\left(-\int_s^\infty [1/w_1(x)]^{q_1} x^{-1} dx\right)$$

$$\leq C\left\{\int_t^\infty K(s,f;A) d\left[-\left(\int_s^\infty [1/w_1(x)]^{q_1} x^{-1} dx\right)^{1/q_1}\right]\right\}^{q_1}$$

$$\leq C\left\{\int_t^\infty [K(s,f;A)/w_1(s)] s^{-1} ds\right\}^{q_1},$$

where the last inequality follows from Proposition 5(vi). Thus

(20)
$$\left\{\int_t^\infty [K(s,f;A)/w_1(s)]^{q_1} s^{-1} ds\right\}^{1/q_1}$$

$$\leq C\int_t^\infty [K(s,f;A)/w_1(s)] s^{-1} ds, \quad 1 < q_1 < \infty.$$

Therefore from (17), (19) and (20) we get

(21)
$$K(\bar{\eta}(t), Tf; B)/\bar{w}_0(\bar{\eta}(t)) \leq C\left\{\int_0^{\eta(t)} [K(s,f;A)/w_0(s)] s^{-1} ds\right.$$

$$\left. + t\int_{\eta(t)}^\infty [K(s,f;A)/w_1(s)] s^{-1} ds\right\}, \quad 1 \leq q_0, q_1 \leq \infty,$$

where the case $q_0 = q_1 = 1$ follows directly from (17) and the case $q_0 = q_1 = \infty$, follows from these two estimations

$$\sup_{0<s<t} K(s,f;A)/w_0(s) = \sup_{0<s<t} [K(s,f;A)/s]s/w_0(s)$$

$$\leq \sup_{0<s<t} [K(s,f;A)/s]\int_0^s [x/w_0(x)] x^{-1} dx$$

$$\leq \sup_{0<s<t} \int_0^s [K(x,f;A)/w_0(x)] x^{-1} dx$$

$$\leq \int_0^t [K(x,f;A)/w_0(x)] x^{-1} dx$$

and

$$\sup_{t<s<\infty} K(s,f;A)/w_1(s) = \sup_{t<s<\infty} K(s,f;A)\int_s^\infty [1/w_1(x)] x^{-1} dx$$

$$\leq \sup_{t<s<\infty} \int_s^\infty [K(x,f;A)/w_1(x)] x^{-1} dx$$

$$\leq \int_t^\infty [K(x,f;A)/w_1(x)] x^{-1} dx.$$

Here, we applied (v) and (vi) of Proposition 5 and the monotonicity properties of the K-functional.

On replacing t by $\bar{\tau}(t)$ in (21) and then on multiplying the resulting inequality by $u(t)\bar{w}_0(t)$ and by integrating, we get

$$\left\{\int_0^\infty [u(t)K(t,Tf;B)]^q dt\right\}^{1/q} \leq C\left\{\int_0^\infty u(t)^q \bar{w}_0(t)^q \left[\int_0^{\bar{\sigma}(t)} [K(s,f;A)/w_0(s)] s^{-1} ds\right.\right.$$

$$+ \bar{\tau}(t) \int_{\bar{\sigma}(t)}^{\infty} \left[K(s,f;A)/w_1(s) \right] s^{-1} ds \Bigg]^q dt \Bigg\}^{1/q},$$

where $\bar{\sigma}(t) = \eta(\bar{\tau}(t))$. From $t\tau'(t)/\tau(t) \geq \alpha > 0$, it follows that $\tau(t) \to 0$ as $t \to 0$; $\tau(t) \to \infty$ as $t \to \infty$ (Remark 8) and similarly, since $t\bar{\tau}'(t)/\bar{\tau}(t) \geq \bar{\alpha} > 0$, $\bar{\tau}(t) \to 0$ as $t \to 0$ and $\bar{\tau}(t) \to \infty$ as $t \to \infty$. The change of variable $\bar{\sigma}(t) = y$ ($t = \sigma(y), dt = \sigma'(y)dy$) on the right side of the above inequality yields

$$\left\{ \int_0^\infty [u(t)K(t,Tf;B)]^q \, dt \right\}^{1/q} \leq C \left\{ \int_0^\infty u(\sigma(y))^q \bar{w}_0(\sigma(y))^q \left[\int_0^y [K(s,f;A)/w_0(s)] s^{-1} ds \right. \right.$$

$$+ \bar{\tau}(\sigma(y)) \int_y^\infty [K(s,f;A)/w_1(s)] s^{-1} ds \Bigg]^q \sigma'(y)dy \Bigg\}^{1/q}.$$

By Minkowski's inequality for $q \geq 1$ and directly if $0 < q < 1$ we obtain

$$\left\{ \int_0^\infty [u(t)K(t,Tf;B)]^q \, dt \right\}^{1/q}$$

$$\leq C \left\{ \left[\int_0^\infty \left[u(\sigma(y))\bar{w}_0(\sigma(y)) \int_0^y [K(s,f;A)/w_0(s)] s^{-1} ds \right]^q \sigma'(y)dy \right]^{1/q} \right.$$

$$+ \left[\int_0^\infty \left[u(\sigma(y))\bar{w}_1(\sigma(y)) \int_y^\infty [K(s,f;A)/w_1(s)] s^{-1} ds \right]^q \sigma'(y)dy \right]^{1/q} \right\}$$

$$\equiv C\{L_0 + L_1\}, \text{ respectively. Here we used the hypothesis } \bar{\tau}(\sigma(t))\bar{w}_0(\sigma(t)) = \bar{w}_1(\sigma(t)) \text{ (recall } \bar{\tau}(t)\bar{w}_0(t) = \bar{w}_1(t)).$$

Now by Lemma 1

$$L_0 \leq C \left(\int_0^\infty [v(t)K(t,f;A)]^p \, dt \right)^{1/p},$$

if and only if

$$\sup_{s>0} \left\{ \int_s^\infty [u(\sigma(y))\bar{w}_0(\sigma(y))]^q \, \sigma'(y)dy \right\}^{1/q} \left\{ \int_0^s [v(t)tw_0(t)]^{-p'} \, dt \right\}^{1/p'} < \infty.$$

Replacing $\sigma(y)$ by t one gets (9).

Similarly,

$$L_1 \leq C \left(\int_0^\infty [v(t)K(t,f;A)]^p \, dt \right)^{1/p},$$

if and only if

$$\sup_{s>0} \left\{ \int_0^s [u(\sigma(y))\bar{w}_1(\sigma(y))]^q \, \sigma'(y)dy \right\}^{1/q} \left\{ \int_s^\infty [v(t)tw_1(t)]^{-p'} \, dt \right\}^{1/p'} < \infty$$

and replacing $\sigma(y)$ by t in the first integral this implies (10).

If $-t\tau'(t)/\tau(t) \geq \alpha > 0$ the result also holds under the obvious modifications. This completes the proof of the theorem.

We now extend Theorem 9 to the case $1 < q < p < \infty$.

Theorem 10. Suppose $A = (A_0, A_1)$, $B = (B_0, B_1)$ are two interpolation couples, $w_i, \bar{w}_i \in B_\psi$, $i = 0, 1$ and $\tau = w_1/w_0$, $\bar{\tau} = \bar{w}_1/\bar{w}_0$ satisfy $|t\tau'(t)/\tau(t)| \geq \alpha > 0$ and

$t\bar\tau'(t)/\bar\tau(t) \ge \bar\alpha > 0$. Further, let $\eta = \tau^{-1}$, $\bar\eta = \bar\tau^{-1}$, $\sigma = \bar\eta \circ \tau$ and T is a quasi-linear operator satisfying $T : A_{w_i,q_i} \to B_{\bar w_i, \bar q_i}$, $i = 0,1$, $1 \le q_i, \bar q_i \le \infty$. If $t\tau'(t)/\tau(t) \ge \alpha > 0$, $1 \le q < p < \infty$ and $1/r = 1/q - 1/p$; u, v non-negative weights satisfying

$$(22) \qquad \left[\int_0^\infty \left[\left(\int_{\sigma(s)}^\infty [u(t)\bar w_0(t)]^q \, dt \right)^{1/q} \left(\int_0^s [v(t)tw_0(t)]^{-p'} \, dt \right)^{1/q'} \right]^r \right.$$
$$\left. \times (v(s)sw_0(s))^{-p'} \, ds \right]^{1/r} < \infty$$

and

$$(23) \qquad \left[\int_0^\infty \left[\left(\int_0^{\sigma(s)} [u(t)\bar w_1(t)]^q \, dt \right)^{1/q} \left(\int_s^\infty [v(t)tw_1(t)]^{-p'} \, dt \right)^{1/q'} \right]^r \right.$$
$$\left. \times (v(s)sw_1(s))^{-p'} \, ds \right]^{1/r} < \infty$$

then (11) holds.

If $-t\tau'(t)/\tau(t) \ge \alpha > 0$, the implication '(22) & (23) $\Rightarrow$ (11)' holds provided the ranges of the first inner integrals of (22) and (23) are changed from $(\sigma(s), \infty)$ to $(0, \sigma(s))$ and from $(0, \sigma(s))$ to $(\sigma(s), \infty)$. (In case $q = 1 < p$, the conditions (22) and (23) take the following forms : if $t\tau'(t)/\tau(t) \ge \alpha > 0$, then

$$\left[\int_0^\infty \left(\int_{\sigma(s)}^\infty [u(t)\bar w_0(t)]^q \, dt \right)^{p'} (v(s)sw_0(s))^{-p'} \, ds \right]^{1/p'} < \infty,$$

and

$$\left[\int_0^\infty \left(\int_0^{\sigma(s)} [u(t)\bar w_1(t)]^q \, dt \right)^{p'} (v(s)sw_1(s))^{-p'} \, ds \right]^{1/p'} < \infty,$$

respectively; If $-t\tau'(t)/\tau(t) \ge \alpha > 0$, then the inner integrals in the two last weight conditions are to be taken from 0 to $\sigma(s)$, respectively from $\sigma(s)$ to ∞).

Proof. Estimate L_0 and L_1 in the proof of Theorem 9 by Lemma 2(a) and (b) respectively.

We now give examples of weights for which Theorems 9 and 10 hold.

Example 11. The weight functions $w_i(t) = t^{\theta_i}$, $\bar w_i(t) = t^{\bar\theta_i}$, $u(t) = t^{-(\bar\theta_0 + \alpha + 1/q)}$, $v(t) = t^{-(\theta_0 + \lambda\alpha/\bar\lambda + 1/p)}$, $0 < \theta_i, \bar\theta_i < 1$, $i = 0, 1$, $\lambda = \theta_1 - \theta_0 > 0$, $\bar\lambda = \bar\theta_1 - \bar\theta_0 > 0$, $0 < \alpha < \bar\lambda$ satisfy the weight conditions (9) and (10) for $1 < p \le q < \infty$, to obtain the classical reiteration theorem for real interpolation spaces.

Examples 12. $w_i(t) = t^{\theta_i}$, $\bar w_i(t) = t^{\bar\theta_i}$, $u(t) = e^{-t}t^{-\bar\theta_0}$, $v(t) = t^{-(1+\theta_0)}e^t$, $0 < \theta_i, \bar\theta_i < 1$, $i = 0, 1$, $\lambda = \theta_1 - \theta_0 < 0$, $\bar\lambda = \bar\theta_1 - \bar\theta_0 > 0$ satisfy (9) and (10) and hence satify Theorem 9.

Examples 13. The weight functions $w_i(t) = t^{\theta_i}$, $\bar w_i(t) = t^{\bar\theta_i}$, $u(t) = t^{-(\bar\theta_0 + \alpha + 1/q)}$, $v(t) = t^{-(1+\theta_0)}e^t$, $0 < \theta_i, \bar\theta_i < 1$, $i = 0, 1$, $0 < \alpha < \min(\bar\lambda/(r\lambda), \bar\lambda)$, $1 < q < p < \infty$, $1/r =$

$1/q - 1/p$, $\lambda = \theta_1 - \theta_0 > 0$, $\bar{\lambda} = \bar{\theta}_1 - \bar{\theta}_0 > 0$ satisfy (22) and (23) and therefore the conclusion of Theorem 10.

Acknowledgements. The author would like to express his gratitude to Prof. H. Heinig of Mathematics and Statistics Department, McMaster University, Ontario, Canada, for suggesting the problem, in weighted inequalities, while supervising the author's doctoral study.

References

1. K.F. Andersen and B. Muckenhoupt; Weighted weak type Hardy inequalities with applications to Hilbert transform and maximal functions; Studia Math. 72 (1982), 9-26.

2. J.S. Bradley; Hardy inequalities with mixed norms; Canad. Math. Bull. 21 (1978), 405-408.

3. A.P. Calderón and R. Scott; Sobolev type inequalities for $p > 0$; Studia Math. 62 (1978), 73-92.

4. J. Gustavsson; A function parameter in connection with interpolation of Banach spaces; Math. Scand. 42 (1978), 289-305.

5. H.P. Heinig; Interpolation of quasi-normed spaces involving weights; Canad. Math. Soc. Conf. Proc. Vol. 1, 1980 Seminar on Harmonic Analysis, 245-267, 1981.

6. H.P. Heinig; Weighted norm inequalities for certain integral operators II; Proc. Amer. Math. Soc. (3) 95 (1985), 387-396.

7. T.F. Kalugina; Interpolation of Banach spaces with functional parameter. The Reiteraction theorem; Vestnik Moskovskoyo University, Ser.1, Math. Mech. 30 (6) (1975), 68-77 (Engl. Transl. Moscow University Math. Bull. 30 (6) (1975), 108-116).

8. W. Mazja; Einbettungassätze für Sobolevsche Räume; Teil 1, Teubner Texte zur Math., Teubner Verl. Leipzig (1979).

9. E. Sawyer; Weighted Lebesgue and Lorentz norm inequalities for the Hardy operators; Trans. A.M.S. 281 (1) (1984), 329-337.

Salah A.A. Emara
Department of Science
Mathematics Unit
The American University in Cairo
113 Sharia Kasr El-Aini
P.O.Box 2511, Cairo, EGYPT.

SOME SINGULAR INTEGRALS
ON THE AFFINE GROUP

G. I. Gaudry*

Mathematics Discipline, Flinders University
PO Box 2100, Adelaide 5001
Australia

Introduction

The affine group of the plane is usually thought of as the set $\{(b, a) : a \in \mathbb{R}^+, b \in \mathbb{R}\}$ with the composition law $(b, a)(d, c) = (b + ad, ac)$. For the sake of convenience, we realise it as $G = \{(s, t) : s, t \in \mathbb{R}\}$, with group product $(u, v)(s, t) = (u + e^v s, v + t)$.

The left-invariant and right-invariant Haar measures on G are, respectively, $dm(s, t) = e^{-t} ds dt$ and $dn(s, t) = ds dt$. They are related by the modular function $\Delta(s, t) = e^{-t}$ via the relationship $dm = \Delta dn$. All of the integrals and function spaces in this paper are taken with respect to left Haar measure, unless specified otherwise.

The group G is a solvable Lie group. Its Lie algebra is $\mathbb{R}^2$, with basis vectors $X = (1, 0)$ and $Y = (0, 1)$. The Lie structure is determined by the specification that $[X, Y] = -X$, and the exponential mapping is given by

$$\exp(u, v) = \left(\frac{e^v - 1}{v} u, v \right).$$

The basis vectors X and Y determine corresponding right-invariant vector fields, also denoted X and Y, such that

$$X f(s, t) = \frac{\partial f}{\partial s}; \qquad Y f(s, t) = s \frac{\partial f}{\partial s} + \frac{\partial f}{\partial t}.$$

The formal Laplacian

$$L_0 = -X^2 - Y^2$$

1980 *Mathematics Subject Classification* (1985 revision). Primary 42B20, 43A80; secondary 22E31.

* Research supported by the Australian Research Council

is, expressed in coordinates,

$$L_0 = -(1+s^2)\frac{\partial^2}{\partial s^2} - 2s\frac{\partial^2}{\partial s\partial t} - \frac{\partial^2}{\partial t^2} - s\frac{\partial}{\partial s}.$$

It can be thought of as having domain $C_c^\infty(G)$.

There are, up to equivalence, just two infinite-dimensional representations σ^+ and σ^-, of G. The are obtained by induction from the abelian normal subgroup $\{(s,0) : s \in \mathbb{R}\}$, act on $L^2(\mathbb{R})$, and are given by

$$[\sigma^\pm(s,t)f](x) = \exp(\pm ise^x)f(x+t).$$

The derived representations are

$$[d\sigma^\pm(X)f](x) - \frac{d}{ds}\exp(\pm ise^x)f(x)|_{s=0} - \pm ie^x f(x),$$

$$[d\sigma^\pm(Y)f](x) = \frac{d}{dt}f(x+t)|_{t=0} = \frac{d}{dx}f(x).$$

Acknowledgement. This paper is an outline of joint work with Peter Sjögren and Tao Qian. Complete proofs and elaborations of some of the remarks made here are to appear in [2].

Let L be the closure of L_0 on $L^2(G, dm)$. The main properties of L are summarised in the following lemma. They can be proved by using [8] and [9].

Lemma 1. (a) *The operator L is positive self-adjoint, one-one, with dense domain and dense range. The inverse also has these properties.*

(b) *If $-1 \le \alpha \le 1$, the fractional power L^α is defined and has the same properties as L and L^{-1} listed in (a).*

The resolvent kernels of L have been calculated by Hulanicki [3].

Lemma 2. *If $\lambda \notin [0, +\infty) = \mathbb{R}^+$, then $(\lambda - L)^{-1}$ is a bounded operator. It is given by*

$$(\lambda - L)^{-1}f(x) = k_\lambda * f(x) = \int_G k_\lambda(y)f(y^{-1}x)dm(y),$$

where

$$k_\lambda(s,t) = e^{t/2}\mathfrak{Q}_{-\nu-1/2}(\cosh t + \frac{1}{2}s^2 e^{-t}), \tag{1}$$

$\nu = \sqrt{\lambda}$, and Q_ρ is the Legendre function of the second kind with parameter $\mu = 0$. *(The square-root is the principal branch, corresponding to slitting the plane along the positive axis.)*

For the main properties of the functions Q_λ see [5]. The kernel k_0 in (2) makes sense, is locally integrable, and although 0 is not in the resolvent set, we have the "inversion formula"

$$k_0 * (-L)\phi = \phi \qquad (\phi \in C_c^\infty(G)).$$

Riesz transforms

In view of Lemma 1, it is possible to consider the operators $XL^{-1/2}$, $L^{-1/2}X$, and the corresponding ones involving Y. These are called the *Riesz transforms*. An alternative definition of Riesz transform was given by Stein [6] in terms of the Poisson semigroup. Let G be a connected, noncompact semisimple Lie group, K a maximal compact subgroup, and $\{X_1, \ldots, X_d\}$ a basis of the Lie algebra of G. Define the Laplacian L to be the closure in $L^2(G, dm)$ of $-X_1^2 - \cdots - X_d^2$. He proved that, if $1 < p < \infty$, the Riesz transforms $X_i L^{-1/2}$ are bounded on the space $L^p(K \backslash G/K, dm)$ of functions in $L^p(G, dm)$ that are K-bi-invariant.

The Riesz transforms are in any case always bounded on all of $L^2(G, dm)$.

Lemma 3. *The Riesz transforms* $XL^{-1/2}, YL^{-1/2}, L^{-1/2}X$ *and* $L^{-1/2}Y$ *are bounded on* $L^2(G, dm)$.

Proof. This follows from the identity

$$\langle X\phi, X\phi \rangle + \langle Y\phi, Y\phi \rangle = \langle L^{1/2}\phi, L^{1/2}\phi \rangle.$$

See also [1] and [7].

Note that it follows from Lemma 1 that the operators listed in the statement of Theorem 1 are bounded on $L^2(G, dm)$.

The singular integrals

This paper deals not with the Riesz transforms *per se* but with certain associated operators that are formally homogeneous of degree 0. They are $Z_1 L^{-1} Z_2$, $Z_1 Z_2 L^{-1}$, and $L^{-1} Z_1 Z_2$, where Z_1 and Z_2 are arbitrary nonzero right-invariant vector fields. The main results are as follows.

Theorem 1. *The operator* $Z_1 L^{-1} Z_2$ *is of weak type (1,1) and of strong type (p,p) if* $1 < p < \infty$.

Theorem 2. *The operators* $Z_1 Z_2 L^{-1}$ *and* $L^{-1} Z_1 Z_2$ *are unbounded on every* $L^p(G, dm)$ $(1 < p < \infty)$ *and are not of weak type (1,1).*

It is worth mentioning that Burns and Robinson [1] have established recently certain other results about "classical singular integrals" on Lie groups. They proved that, in the notation introduced above, if G is an arbitrary connected Lie group, n is a positive integer, and $\lambda > 0$, then for every $p \in (1, +\infty)$ there is a constant $C_{p,n,\lambda}$ such that

$$\|X_{i_1} \cdots X_{i_n} f\|_p \leq C_{p,n,\lambda} \|(\lambda + L)^{n/2} f\|_p$$

for all $f \in C_c^\infty(G)$. The constant $C_{p,n,\lambda}$ tends to $+\infty$ as $\lambda \to 0$.

The kernels

Each of the operators listed in Theorems 1 and 2 is a principal value convolution operator. Here is a list of the operators formed out of the basis vectors X and Y, along with their kernels. The "$\sim$" sign means that to within a constant operator, the operator is given by principal value convolution with the stated kernel: we shall say, for the sake of simplicity, that the kernel is the kernel of the given operator. The terms in boxes are locally integrable. These formulas are derived essentially by integration by parts.

Lemma 4. If we write $z = \cosh t + \frac{1}{2}s^2 e^{-t}$, we have

$$XL^{-1}X \sim e^t \frac{\partial^2}{\partial s^2} k_0(s,t) = e^{\frac{t}{2}}\mathfrak{Q}'_{-\frac{1}{2}}(z) + s^2 e^{-\frac{t}{2}}\mathfrak{Q}''_{-\frac{1}{2}}(z)$$

$$YL^{-1}Y \sim \left(\boxed{s\frac{\partial^2}{\partial s\partial t}} - \boxed{s\frac{\partial}{\partial s}} + \frac{\partial^2}{\partial t^2} - \boxed{\frac{\partial}{\partial t}}\right)k_0(s,t)$$

$$= e^{\frac{t}{2}}\left[(-\frac{1}{4})\mathfrak{Q}_{-\frac{1}{2}}(z) + (\cosh t - s^2 e^{-t})\mathfrak{Q}'_{-\frac{1}{2}}(z)\right.$$
$$\left. + (\sinh^2 t - (1/4)s^4 e^{-2t})\mathfrak{Q}''_{-\frac{1}{2}}(z)\right]$$

$$XL^{-1}Y \sim \left(\frac{\partial^2}{\partial s\partial t} - \boxed{\frac{\partial}{\partial s}}\right)k_0(s,t)$$

$$= -\frac{3}{2}se^{-\frac{t}{2}}\mathfrak{Q}'_{-\frac{1}{2}}(z) + se^{-\frac{t}{2}}(\sinh t - \frac{1}{2}s^2 e^{-t})\mathfrak{Q}''_{-\frac{1}{2}}(z)$$

$$YL^{-1}X \sim e^t\left(\boxed{s\frac{\partial^2}{\partial s^2}} + \frac{\partial^2}{\partial s\partial t} + \boxed{\frac{\partial}{\partial s}}\right)k_0(s,t)$$

$$= \frac{3}{2}se^{\frac{t}{2}}\mathfrak{Q}'_{-\frac{1}{2}}(z) + se^{\frac{t}{2}}(\sinh t + \frac{1}{2}s^2 e^{-t})\mathfrak{Q}''_{-\frac{1}{2}}(z)$$

$$X^2 L^{-1} \sim \frac{\partial^2}{\partial s^2}k_0(s,t)$$
$$\sim e^{-t}XL^{-1}X$$

$$XYL^{-1} \sim \left(\boxed{s\frac{\partial^2}{\partial s^2}} + \frac{\partial^2}{\partial s\partial t} + \boxed{\frac{\partial}{\partial s}}\right)k_0(s,t)$$
$$\sim e^{-t}YL^{-1}X$$

$$YXL^{-1} \sim \left(\boxed{s\frac{\partial^2}{\partial s^2}} + \frac{\partial^2}{\partial s\partial t}\right)k_0(s,t)$$

$$L^{-1}X^2 \sim e^{2t}\frac{\partial^2}{\partial s^2}k_0(s,t)$$
$$\sim e^t XL^{-1}X$$

$$L^{-1}XY \sim e^t \frac{\partial^2}{\partial s\partial t}k_0(s,t)$$

$$L^{-1}YX \sim e^t\left(\frac{\partial^2}{\partial s\partial t} - \boxed{\frac{\partial}{\partial s}}\right)k_0(s,t)$$
$$\sim e^t XL^{-1}Y.$$

Behaviour at infinity

The principal observation here is that the kernels of the "two-sided" operators $XL^{-1}X$, $XL^{-1}Y$, $YL^{-1}X$, and $YL^{-1}Y$ are all integrable at infinity. This follows from the asymptotic formulas at ∞ that can be obtained from [MOS], pp. 174, 153:

$$\mathfrak{Q}_{-\frac{1}{2}}(z) = 2C_1 z^{-\frac{1}{2}} + \text{integrable term}, \qquad \mathfrak{Q}'_{-\frac{1}{2}}(z) = -C_1 z^{-3/2} + \text{integrable term} \quad (2)$$

and

$$\mathfrak{Q}''_{-\frac{1}{2}}(z) = \frac{3}{2} C_1 z^{-5/2} + \text{integrable term}. \tag{3}$$

The further analysis of our operators proceeds by breaking each kernel into its *local part* and its *part at infinity*. Fix a function $\psi \in C_c^\infty(G)$ that is 1 on a neighbourhood of the identity. The local part is $k = K\psi$ where K is the kernel of the given operator. The part at infinity is $K - k$. Since the kernels are locally integrable on the complement of the origin, the mapping properties of the local part and of the part at infinity are independent of the choice of the function ψ. By taking into consideration the boxed terms in Lemma 4 and the fact that the two-sided operators have kernels that are integrable at infinity, we obtain the following.

Lemma 5. *The local parts of all of the operators (two-sided or otherwise) are bounded on* $L^2(G, dm)$.

Suppose now that the operator A is given by $Af = k * f$, where k has compact support. The essence of the following lemma is that if A is bounded on $L^2(G, dm)$ and satisfies estimates of Calderón-Zygmund type, then A is of weak type $(1, 1)$. Cf also [1].

Lemma 6. *Let* $L(x, y) = k(xy^{-1})$, $(x, y) \in \mathbb{R}^2$. *Suppose that* $L \in C^\infty$ *for* $x \neq y$ *and that*

$$|L(x, y)| + |x - y|(|\nabla_x L| + |\nabla_y L|) \le C|x - y|^{-2}$$

for all $x \neq y$. *If* A *is bounded on* $L^2(G, dm)$, *then* A *is of weak type* $(1, 1)$.

Proof. This has two principal ingredients. First, the group G is locally Euclidean; so if h is any function in $C_c^\infty(G)$, then the operator $A_h : f \longmapsto \int L(x, y) h(y) f(y)\, dy$ is of weak type $(1, 1)$ from $L_C^1(\mathbb{R}^2)$ to $L^1(\mathbb{R}^2)$ for every compact set C. The local weak-type behaviour of the operators A_h is then pieced together by using the following covering lemma. (The affine group has a natural left-invariant hyperbolic metric, obtained, for instance, by identifying it with the symmetric space $SL(2, \mathbb{R})/SO(2)$.)

Lemma 7. *Let* H_ρ *denote the hyperbolic ball of radius* ρ *centered at* e. *Given* $\varepsilon > 0$, *and* $\delta > 0$, *there exist a sequence* $t_1, t_2, \ldots$ *of points of* G *and* $r > 0$, $s > 0$ *such that*
  (i) $G = \bigcup_i H_\varepsilon t_i$.
  (ii) *Each* $y \in G$ *belongs to at most* r *of the sets* $H_\varepsilon t_i$.
  (iii) *Each* $y \in G$ *belongs to at most* s *of the sets* $H_{\varepsilon + \delta} t_i$.

We can now deduce Theorem 1, since we have the asymptotic formulas, as $z \searrow 1$, that follow from [MOS], pp. 174, 196:

$$\mathfrak{Q}_{-\frac{1}{2}}^{(m)}(z) = (-1)^m \frac{(m - 1)!}{2} (z - 1)^{-m} + O((z - 1)^{-m+1}), \qquad m = 1, 2, \ldots.$$

It is therefore not hard to check that the conditions of Lemma 7 are satisfied by the local parts of the kernels of the two-sided operators; and the parts at infinity are in any case integrable.

Behaviour on L^p of the parts of the other kernels at infinity

It follows from Lemma 4 that (to within a multiple of the Dirac delta at e and terms that are integrable) the kernels of the operators $Z_1 Z_2 L^{-1}$ behave at infinity like

$$A(e^{-\frac{t}{2}}\mathfrak{Q}'_{-\frac{1}{2}}(z) + s^2 e^{-\frac{3t}{2}}\mathfrak{Q}''_{-\frac{1}{2}}(z)) + B(se^{-\frac{3t}{2}}\mathfrak{Q}''_{-\frac{1}{2}}(z)) + Cse^{-\frac{t}{2}}\mathfrak{Q}'_{-\frac{1}{2}}(z), \qquad (4)$$

where at least one of A, B and C is not zero. The expression in (4) is integrable in the region $t \geq 0$. For large negative t it behaves like $r(s)e^t$, where

$$\frac{a(2s^2 - 1)}{(1 + s^2)^{\frac{5}{2}}} + \frac{bs}{(1 + s^2)^{\frac{5}{2}}} + \frac{cs}{(1 + s^2)^{\frac{3}{2}}}$$

and $r \neq 0$.

Consider then the kernel

$$W(s,t) = r(s)e^t \chi_{(-\infty,0]}(t) \qquad (5)$$

and the family of test functions $f_R = \chi_{E_R}$, where $E_R = \{(s,t) : 0 < s < e^t, \, 0 < t < R\}$. It is not hard to check that $\|f_R\|_p$ behaves like $R^{1/p}$ and that $\|W * f_R\|_p$ grows at least as fast as R as $R \to \infty$. This confirms the unboundedness of the operators $Z_1 Z_2 L^{-1}$. The operators $L^{-1}Z_1 Z_2$ are conjugate to these. They are therefore also unbounded.

Weak (1, 1) behaviour of the part at infinity

The operator $L^{-1}Z_1 Z_2$ can be shown not to be of weak type $(1,1)$ relatively easily. One simply proves that the part of the kernel at infinity is not in weak-L^1. This relies again on the asymptotics in (2) and (3).

The operator $Z_1 Z_2 L^{-1}$ is more difficult. Refer again to the analysis in the preceding section. If $f \geq 0$, $f(s,t) = 0$ for $t \leq 0$, and W is given by (5), then

$$W * f(s,t) = r_{e^{-t}} *_{\mathbb{R}} \check{h}(se^{-t}), \qquad t < 0,$$

where

$$h(u) = \int_{-\infty}^{\infty} f(-e^{-v}u, -v)dv.$$

Now the mapping $f \mapsto h$ is in fact surjective, and if $h \geq 0$ is given, it is possible to choose f so that f maps to h and $\|f\| = \|h\|$. In order to show that the operator $Z_1 Z_2 L^{-1}$ is unbounded, it is therefore enough to disprove the inequality

$$\mu(\{(x,y) : y > 1, |r_y * h(x)| > \lambda\}) \leq \frac{C}{\lambda}\|h\|_{L^1(\mathbb{R})}, \qquad (6)$$

where $d\mu = dx\,dy/y$. In fact it is possible to dispense with the condition $y > 1$ in (6).

Outline of a counter-example to (6).

The function r has the properties

$$r(s) = O(|s|^{-2}), \qquad r'(s) = O(|s|^{-3}) \quad (|s| \to +\infty), \qquad \int_{-\infty}^{\infty} r(s)\, ds = 0.$$

Choose $\phi \in C^\infty(\mathbb{R})$ with the same properties as r and so that $\int_{-\infty}^\infty r(s)\phi(-s)\,ds \neq 0$. This implies that $r_y * \phi(x) > \delta$ on some set U of positive measure in $[-1,1] \times [\frac{1}{2}, 2]$.

Fix large positive integers p, q and N, and let

$$h_N = \sum_{n=n_0}^{N} \sum_{k=1}^{2^{qn/p}} \pm \phi(2^{qn}x - kp)$$

$$= \sum_{n=n_0}^{N} \sum_{k=1}^{2^{qn/p}} \pm \phi_{nk}$$

where n_0 is the smallest n for which $2^{qn} \geq p$. Notice that ϕ_{nk} involves translates by $p/2^{qn}$ of (increasingly lacunary) dilates of ϕ. Notice also that, if ϕ is concentrated near 0, then ϕ_{nk} is concentrated near $kp/2^{qn}$, and that for fixed n, the functions ϕ_{nk} may be thought of as having close-to-disjoint supports. For fixed k, the main parts of the supports of the functions ϕ_{nk} shrink in a lacunary manner as $n \nearrow +\infty$. Now

$$r_y * \phi_{nk}(x) = r_{2^{qn}y} * \phi(2^{qn}x - pk) > \delta$$

on $U_{nk} \subseteq [-2^{-qn} + kp2^{-qn}, 2^{-qn} + kp2^{-qn}] \times [2^{-qn-1}, 2^{-qn+1}]$. The sets U_{nk} are pairwise disjoint.

The remaining part of the argument involves proving the following two statements.

Claim 1. *One can choose q, p and $\lambda > 0$ so that, for all choices of $\pm$ signs, and all large N,*

$$\mu(\{(x,y) : y > 0, |r_y * h_N(x)| > \lambda\}) \geq CN,$$

where $C > 0$ is independent of the choice of signs and of N.

Claim 2. *The signs $\pm$ can be chosen so that $\|h_N\|_{L^1(\mathbb{R})} = O(\sqrt{N})$, as $N \to \infty$.*

In the proof of Claim 2, randomisation arguments are used to reduce the estimation of $\|h_N\|_{L^1(\mathbb{R})}$ to estimating

$$\int_{-\infty}^\infty \left(\sum_{n=n_0}^{N} \sum_{k=1}^{2^{qn/p}} |\phi_{nk}|^2 \right)^{\frac{1}{2}} dx.$$

This can be handled by using the decay properties of ϕ and ϕ'. The details are lengthy and technical, and will be presented in [2].

References

[1] R. J. Burns and D. W. Robinson, *A priori inequalities on Lie groups*. Preprint.

[2] G. I. Gaudry, P. Sjögren, T. Qian, *Singular integrals associated to the Laplacian on the affine group ax + b*. Submitted.

[3] A. Hulanicki, *On the spectrum of the Laplacian on the affine group of the real line*. Studia Math. **LIV** (1976), 199–204.

[4] T. Kato, "Perturbation theory for linear operators". Springer-Verlag, Berlin· Heidelberg ·New York 1966.

[5] W. Magnus, F. Oberhettinger, R. P. Soni, "Formulas and Theorems for the Special Functions of Mathematical Physics". Springer-Verlag, Berlin·Heidelberg·New York 1966.

[6] E. M. Stein, *Topics in harmonic analysis related to the Littlewood-Paley theory*. Annals of Math. Studies, No. 63. Princeton University Press. Princeton, NJ 1970.

[7] R. S. Strichartz, *Analysis of the Laplacian on a complete Riemannian manifold*. J. Funct. Anal. **52** (1983), 48–79.

[8] H. Tanabe, "Equations of Evolution". Pitman, London·San Francisco·Melbourne 1979.

[9] K. Yosida, "Functional Analysis". Second Edition. Springer-Verlag, Berlin·Heidelberg ·New York 1968.

FROM RIESZ PRODUCTS TO RANDOM SETS

Jean-Pierre KAHANE

Université de Paris-Sud
Mathématiques
Bâtiment 425
91405 ORSAY Cedex

This talk consists of three interrelated parts : I. Riesz products ; II. Random operators coming from multiplication of independent factors ; III. Random coverings. The linkage is the role of singular measures and their dimensional analysis.

I. Riesz products

In 1918, F. Riesz introduced the infinite products

$$\prod_1^\infty (1 + \cos 4^m t)$$

and

$$\prod_1^\infty (1 + a_m \cos 4^m t) \tag{1}$$

$(-1 \le a_m \le 1,\ (a_m)$ dense on $[-1,1])$ in order to give examples of continuous measures whose Fourier coefficients do not tend to zero [25]. Actually F. Riesz used the terminology of functions with bounded variation and considered integrals of the form

$$\int_0^t \prod_1^\infty (1 + a_m \cos 4^m t)\, dt.$$

The corresponding measures can be denoted by

$$\prod_1^\infty (1 + a_m \cos 4^m t)\, dt$$

as well as (1). The main point is that all factors are ≥ 0 and no interference appears when the partial products are developed, in particular the constant term is 1.

The same holds for products of the form

$$\prod_1^\infty \left(1 + \mathrm{Re}(a_j e^{i\lambda_j t})\right) \tag{2}$$

or

$$\prod_1^\infty (1 + \rho_j \cos(\lambda_j t + \varphi_j))$$

where $a_j = \rho_j e^{i\varphi_j}$ are complex numbers of moduli ≤ 1 and the λ_j are positive integers such that

$$\lambda_{j+1}/\lambda_j \geq 3 \quad (j = 1, 2, \cdots \tag{3}$$

From now on (2) is what we call Riesz products.

Suppose that the λ_j are fixed. We write a for the sequence (a_j) and μ_a for the corresponding measure. What is the relative behavior of two such measures, μ_a and μ_b ? It is guessed, not yet proved, that they are either mutually singular (= orthogonal) or mutually absolutely continuous (= equivalent).

The first result in that direction was obtained by J. Peyrière in 1973 [22] and independently by G. Brown and W. Moran [2] :

$$\sum_1^\infty |a_j - b_j|^2 = \infty \Rightarrow \mu_a \perp \mu_b \tag{4}$$

$$\left\{ \begin{array}{l} \sum |a_j - b_j|^2 < \infty \\ \sup |a_j| < 1 \end{array} \right\} \Rightarrow \mu_a \sim \mu_b. \tag{5}$$

The proof of Peyrière is simple and beautiful and I shall give it later.

However it is not true that $\sum |a_j - b_j|^2 < \infty$ implies $\mu_a \sim \mu_b$. In 1988 [18] Kilmer and Saeki discovered the role of another distance in the disc $\{z = re^{i\theta}, 0 \leq r \leq 1\}$, given by the metric

$$ds^2 = d\theta^2 + (1 - r)^{-1/2} dr^2.$$

The squared distance between a and b in this metric, $d^2(a, b)$, is equivalent to

$$\rho^2(\sin^2\theta + (1 - r)^{-1/2}\cos^2\theta)$$

when $a = (r + \rho e^{i\theta})e^{i\omega}$, $b = (r - \rho e^{i\theta})e^{i\omega}$. Here is the guess :

$$\sum_1^\infty d^2(a_j, b_j) = \infty \Rightarrow \mu_a \perp \mu_b \tag{6}$$

$$\sum_1^\infty d^2(a_j, b_j) < \infty \Rightarrow \mu_a \sim \mu_b. \tag{7}$$

Kilmer and Saeki proved that the assumptions in (6) and (7) imply the conclusions *almost surely* if the a_j and b_j are multiplied randomly by $e^{i\omega_j}$ (independent random variables equally distributed on the circle). As a consequence, (6) and (7) hold when (3) is replaced by a stronger lacunarity condition, such as

$$\text{(3) and } \sum_{1}^{\infty} \left(\frac{\lambda_j}{\lambda_{j+1}}\right)^2 < \infty.$$

This can be seen by convolution with a measure of the type

$$\prod_{1}^{\infty} \gamma_{n_j}(\lambda_j t - \omega_j)$$

where

$$\gamma_n(t) = \gamma_n(-t) = \sum_{-n}^{n} \hat{\gamma}_n(m) e^{imt} \geq 0$$

$$\hat{\gamma}_n(0) = 1, \quad |\hat{\gamma}_n(1) - 1| \leq \frac{C}{n^2}$$

$$n_j \lambda_j < \frac{1}{2}\lambda_{j+1}.$$

The conjecture is supported by a result of F. Parreau (1989) [21] obtained in a quite different way : (6) holds when $|a_j| = |b_j|$ (in this case, (4) and (6) are the same).

Let me give the proofs of (4) and (5).

(4) relies on the fact that the functions $e^{i\lambda_j t} - \frac{1}{2}\bar{a}_j$ are orthogonal in $L^2(\mu_a)$. Therefore, given $(\alpha_j) \in \ell^2$, the series $\sum \alpha_j \left(e^{i\lambda_j t} - \frac{1}{2}\bar{a}_j\right)$ converges in $L^2(\mu_a)$. Suppose μ_a non orthogonal to μ_b. Then both series $\sum \alpha_j \left(e^{i\lambda_j t} - \frac{1}{2}\bar{a}_j\right)$ and $\sum \alpha_j \left(e^{i\lambda_j t} - \frac{1}{2}\bar{b}_j\right)$ converge in $L^2(\mu_a \wedge \mu_b)$. Taking differences this implies $(a_j - b_j) \in \ell^2$.

(5) comes from

$$\prod \frac{1 + \text{Re}(a_j e^{i\lambda_j t})}{1 + \text{Re}(b_j e^{i\lambda_j t})} \sim \exp \sum \text{Re}\left((\alpha_j - b_j)e^{i\lambda_j t}\right)$$

under the essential assumption $\| a \|_\infty < 1$ together with $a - b \in \ell^2$.

From these results and proofs we can formulate three points.

α) Riesz products are a good way to construct singular measures ;
β) Riesz products are a good way to appreciate the behavior of lacunary trigonometric series ;

γ) Riesz products are a good substitute for exponentials of lacunary trigonometric series.

Let me add a few comments on each of these points.

α) The dimensional analysis of Riesz products.
Let us consider the case $a_j = a$, $\lambda_j = e^{\xi j}$, and write

$$\sigma(a,b) = \int \log(1 + a\cos x)\mu_b(dx) \quad (-1 \le a, b \le 1).$$

Here is an interesting formula involving three measures μ_a, μ_b, μ_c $(-1 \le a, b, c \le 1)$, where ξ is fixed, and $I_\epsilon(x)$ denotes the interval $(x - \epsilon, x + \epsilon)$:

$$\lim_{\epsilon \to 0} \frac{\log \mu_a(I_\epsilon(x))}{\log \mu_b(I_\epsilon(x))} = \frac{1 - \xi\sigma(a,c)}{1 - \xi\sigma(b,c)} \qquad \mu_c\text{--a.e.} \tag{8}$$

(Fan Ai-Hua, unpublished). When $b = 0$, then $\mu_b = \lambda$, the normalized Lebesgue measure. When moreover $c = 0$ we have a formula of Peyrière [23]

$$\lim_{\epsilon \to 0} \frac{\log \mu_a(I_\epsilon(x))}{\log \lambda(I_\epsilon(x))} = 1 - \xi\sigma(a,a) \qquad \mu_a\text{--a.e.} \tag{9}$$

which implies that μ_a is unidimensional and that the dimension of μ_a is $1 - \xi\sigma(a,a)$. Note that $\lambda(I_\epsilon(x)) = 2\epsilon$.

Generally speaking, a measure μ is unidimensional if, for some α, μ is carried by a Borel set of dimension α and if it kills all Borel sets of dimension $< \alpha$. Then α is the dimension of μ. Here is an equivalent definition (Fan Ai-Hua, thesis [6]) :

$$\underline{D}(\mu, x) = \alpha \qquad \mu - \text{a. e.}$$

where

$$\underline{D}(\mu, x) = \liminf_{\epsilon \to 0} \frac{\log \mu(I_\epsilon(x))}{\log \lambda(I_\epsilon(x))}.$$

In other words, μ is $(\alpha+)-$Hölder μ almost everywhere, and nothing better.

However we can consider the Borel sets

$$E_\beta = \{x ; \underline{D}(x, \mu) = \beta\} \tag{10}$$

and their dimensions

$$\varphi_\mu(\beta) = \dim E_\beta. \tag{11}$$

The study of $\varphi_\mu(\beta)$ is the multifractal analysis of unidimensional measures, a domain in which physicists and mathematicians compete (P. Collet, J. Peyrière, G. Brown, G. Michon, in particular on Riesz products).

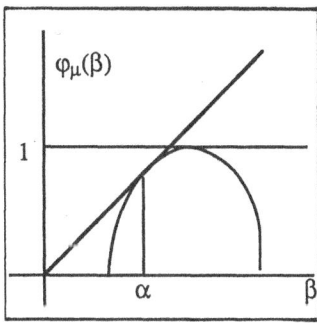

The first case when $\varphi_\mu(\beta)$ was computed is for an analogue of Riesz products, namely

$$\prod_1^\infty (1 + ar_j(t)) \tag{12}$$

where the $r_j(\cdot)$ are the Rademacher functions. This measure has been considered for a long time (see for example Salem [26]) but its multifractal analysis is quite recent.

β) J. Peyrière [24] and Fan Ai-Hua [8] proved that the series

$$\sum_1^\infty \alpha_j \left(e^{i\lambda_j t} - \frac{1}{2}\bar{a}_j \right)$$

converges μ_a-a. e. whenever $(\alpha_j) \in \ell^2$ (here again μ_a is the measure (2)). Moreover the $L^p(\mu_a)$ norms of the maximal function of the partial sums are $O(p^{3/2}(p-1)^{-1})$ (Fan Ai-Hua).

The method of Fan Ai-Hua uses a randomisation of the phases φ_j as a step towards the deterministic result.

γ) There is much in common between lacunary trigonometric series and random trigonometric series, for example gaussian series

$$\sum_{m=0}^\infty r_m(\xi_m \cos mt + \xi'_m \sin mt) \tag{13}$$

where the ξ_m and ξ_m' are independent normalized gaussian variables. (13) converges a. s. a. e. if $(r_m) \in \ell^2$, then exponentiation makes no problem. What when $(r_m) \notin \ell^2$? By analogy with Riesz products we may suspect that $r_m = \frac{a}{\sqrt{m}}$ is a critical case for some a (because the energy in exponentially increasing strips of frequences is constant then, as it is for series

$$\sum a \cos 4^j t$$

associated with Riesz products

$$\prod (1 + a \cos 4^j t)).$$

II. Random operators coming from multiplication of independent factors.

Here is a general frame. We are given a probability space $(\Omega, \mathcal{A}, P)$, a compact (sometimes locally compact) space T, and a sequence of positive independent functions $F_n(t, \omega)$ $(\omega \in \Omega, t \in T, n \in \mathbb{N}^+)$ which we write simply $F_n(t)$. We suppose $EF_n(t) < \infty$ for each t and we write

$$P_n(t) = F_n(t)/EF_n(t)$$

in such a way that $EP_n(t) = 1$, and

$$Q_n(t) = P_1(t)P_2(t)\cdots P_n(t)$$

in such a way that $Q_n(t)$ is a positive martingale for each given t.

Now we consider a positive measure $\sigma \in M^+(T)$. It is easily seen that the random measure $Q_n\sigma$ converges weakly a. s. to a random measure S. We write

$$S = Q\sigma. \tag{14}$$

Q is an operator which transform a measure σ into a random measure S. If $S = 0$ we say that σ is Q-singular. If $ES = \sigma$ we say that σ is Q-regular. It can be checked that σ is decomposed in a unique way in the form $\sigma_r + \sigma_s$, where σ_r is Q-regular and σ_s is Q-singular. Actually $\sigma_r = ES$, $\sigma_s = \sigma - ES$, and both mappings $\sigma \to \sigma_r$, $\sigma \to \sigma_s$ are projections.

Given the sequence $(F_n(t))$, or $(P_n(t))$, we can try to define the Q-regular and the Q-singular measures. Or, given the sequence $(F_n(t))$ and a measure σ, we can investigate the properties of S.

It is rather rare to obtain a complete answer to the first question ; however, it is what we shall do in section III in a particular case. For the second question here is a vague but rather general statement : if σ is unidimensional, $Q\sigma$ is unidimensional a. s., and

$$\begin{array}{llll} \dim Q\sigma &= \dim \sigma - \epsilon(Q) & \text{if} & \dim \sigma \geq \epsilon(Q) \\ Q\sigma &= 0 & & \text{if} & \dim \sigma < \epsilon(Q) \end{array} \right\} \tag{15}$$

where $\epsilon(Q)$ is a positive number, depending on Q, which we call the entropy of Q.

Here is another general remark. Given the sequence $(F_n(t))$ and assuming $EF_n^h(t) < \infty$ for all $h > 0$, we can consider $(F_n(t))$ for, say, countably many h, and obtain a family of operators $Q^{(h)}$. Writing $S^{(h)} = Q^{(h)}\sigma$ one can investigate the mutual properties of these random measures. For example, it is possible to get formulas of the type (8) when we consider 3 values $h = a, b, c$ [14] (actually, these formulas were the motivation for (8)). Is it possible to define a. s. the family $(Q^{(h)})_{h \in \mathbb{R}^+}$? Except for very particular cases (such as in section III) this question seems open.

There are many methods to investigate the properties of S [11][12][6]. Here is the simplest. Consider the martingale $S_n(T)$, where $S_n = Q_n\sigma$. It converges in $L^2(\Omega)$ if and only if $ES_n^2(T) = O(1)$, that is

$$\int\int E(Q_n(t)Q_n(s))\sigma(dt)\sigma(ds) = O(1). \tag{16}$$

Usually (for example when $E(P_n(t)P_n(s)) \geq 1$ for all t, s, n) (16) can be written as

$$I_k(\sigma) = \int\int k(t, s)\sigma(dt)\sigma(ds) < \infty \tag{17}$$

for some kernel k ; $I_k(\sigma)$ is the energy integral. Suppose moreover $I_{kk_1}(\sigma) < \infty$, where kk_1 is the product of k and another positive kernel k_1. Then $EI_{k_1}(S) < \infty$ (easy calculation), therefore $I_{k_1}(S) < \infty$ a. s.

Similar computations, only more complicated, can be done when replacing $L^2(\Omega)$ by $L^p(\Omega)$ where p is an integer ≥ 2 ([11], [5]).

To go further we have to be more specific.

Let us consider independent centered gaussian processes $X_n(t)$ $(n \in \mathbb{N}^+)$. Their law is well defined by the correlation functions

$$p_n(t, s) = E X_n(t)X_n(s).$$

For technical reasons let us suppose $p_n(t, s) \geq 0$ for all t, s, n. Let us consider $F_n(t) = \exp X_n(t)$. Then $P_n(t) = \exp(X_n(t) - 1/2p(t, t))$,

$$E(P_n(t)P_n(s)) = e^{p_n(t,s)}$$

and (17) can be written as

$$\left.\begin{array}{c}\int\int e^{q(t,s)}\sigma(dt)\sigma(ds) < \infty \\ q(t, s) = \sum_1^\infty p_n(t, s)\end{array}\right\}. \tag{18}$$

Since the $p_n(t,s)$ are of positive type, we say that $q(t,s)$ is of sigma positive type. One may think of the case $q(t,s) \simeq u \log \frac{1}{\|t-s\|}$ when $t \in \mathbb{R}^d$, $s \in \mathbb{R}^d$ and $\| t - s \| < \frac{1}{2}$, u being a positive parameter.

(18) expresses a necessary and sufficient condition for σ to be Q-regular <u>and</u> $Q\sigma(T) \in L^2(\Omega)$. One may ask for conditions which imply that σ is Q-regular and $Q\sigma(T) \in L^p(T)$. Here is such a condition, when p is an integer ≥ 2 :

$$\sup_s \int e^{\frac{p}{2}q(t,s)} \sigma(dt) < \infty \qquad [5].$$

This suggests a definition and a conjecture. Given a kernel $k(s,t)$, we say that σ is the k-regular, or regular with respect to k, if σ is a countable union of measures of finite k-energy.

Conjecture A. 1) σ is Q-regular if and only if it regular with respect to the kernel $\exp \frac{1}{2}q(t,s)$; 2) moreover, σ can be written as $\sigma = \sum_1^\infty \sigma_n$ with $Q\sigma_n(T) \in L^p(T)$ if and only if σ is regular with respect to the kernel $\exp \frac{p}{2}q(t,s)$ $(p \geq 1)$.

Here is an important fact, not so easy to prove : the law of Q depends only on the kernel q. Moreover, if we have two kernels q_1, q_2 such that $q_1 \leq q_2$ and we consider the corresponding operators Q_1, Q_2, then "Q_1 is better than Q_2" in every reasonable meaning : σ is Q_1-regular whenever it is Q_2-regular ; $Q_1\sigma$ is α-Hölder, or $Q_1\sigma(T) \in L^p(\Omega)$, whenever the same is true for $Q_2\sigma$; $\epsilon(Q_1) \leq \epsilon(Q_2)$ etc. [11].

Let us consider the particular case $T = \mathbb{R}^d$, $q(t,s) \simeq u \log \frac{1}{\|t-s\|}$ when $\| t - s \|$ is small enough. Then we have (15) with

$$\epsilon(Q) = \frac{u}{2} \qquad (19)$$

when σ is equivalent with the Lebesgue measure λ [11]. In particular, $Q\lambda = 0$ when $u > 2d$ and $EQ\lambda = \lambda$ when $u < 2d$. This proves a conjecture of B. Mandelbrot [19].

Conjecture B. In the particular case $T = \mathbb{R}^d$, $q(t,s) = u \log \frac{1}{\|t-s\|}$ when $\| t - s \|$ is small enough, (15) and (19) hold whatever $\sigma \in M^+(T)$.

Conjecture A implies conjecture B and it is much more ventured than conjecture B.

Instead of gaussian processes $X_n(t)$ one can consider Lévy processes, with the assumption that $E \exp X_n(t) < \infty$. This was done by Fan Ai-Hua ([6], [7]).

To conclude this section let us make three observations.

1. The first example we encountered of independent weights $P_j(t)$ was $P_j(t) = 1 + \rho_j \cos(\lambda_j t + \omega_j)$, then

$$Q\lambda = \prod (1 + \rho_j \cos(\lambda_j t + \omega_j)) \, dt$$

as in the theorem of Kilmer and Saeki (see (6) and (7) and the work of Fan Ai-Hua [8]).

2. Let us go back to the gaussian trigonometric series (13) with $r_m = \frac{a}{\sqrt{m}}\ (m \geq 1)$. Then, formally,

$$q(t,s) = r_0^2 + a^2 \sum_1^\infty \frac{1}{m} \cos m(t-s) \simeq a^2 \log \frac{1}{|t-s|}$$

$$\epsilon(Q) = \frac{a^2}{2}$$

(19)

and the critical value for a is $\sqrt{2}$. Here is the exact meaning of (19) : whenever we decompose the series $r_0^2 + a^2 \sum_1^\infty \frac{1}{m} \cos mt$ in the form $\sum_1^\infty p_n(t)$, where $p_n(t) \geq 0$ and the Fourier coefficients of p_n are ≥ 0, $q(t,s)$ is the kernel of sigma positive type equal to $\sum_1^\infty p_n(t-s)$.

3. The first and still best example of random measures S generated by a random operator Q is the model introduced by B. Mandelbrot [20] and studied by J. Peyrière and myself [17].

III. Random coverings

Let us consider the case $F_n^2(\cdot) = F_n(\cdot)$, that is

$$F_n(t) = 1_{F_n}(t)$$

where the F_n are independent random subsets of T. It is convenient to consider the $G_n = T \backslash F_n$ and to write

$$G_n(t) = 1_{G_n}(t) = 1 - F_n(t).$$

(21)

Usually the G_n are open sets. Then

$$P_n(t) = \frac{1 - G_n(t)}{1 - P(t \in G_n)}$$

and

$$Q_n(t) = 0 \Leftrightarrow t \in G_1 \cup G_2 \cup \cdots \cup G_n.$$

If we have (16), that is

$$\left. \begin{array}{l} \displaystyle\int\int k_N(t,s)\,d\sigma(t)\,d\sigma(s) = O(1) \\[2mm] \displaystyle k_N(t,s) = \prod_{n=1}^N \frac{E\left((1 - G_n(t))(1 - G_n(s))\right)}{(1 - P(t \in G_n))(1 - P(s \in G_n))} \end{array} \right\}$$

(22)

for some $\sigma \in M^+(T)$, carried by K, a compact subset of T, clearly

$$P(\exists t \in K \ : \ t \notin \bigcup_1^\infty G_n) > 0. \tag{23}$$

When (23) holds we say that K is not covered. In the opposite situation

$$P(K \subset \bigcup_1^\infty G_n) = 1$$

we say that K is covered. The problem of covering the circle by random intervals of given lengths ℓ_n was introduced by Dvoretzky [3], studied by Billard [1] and by myself (see references in [10]), solved by Shepp [27], generalized to the covering of compact metric spaces by random balls by Hoffman-Jørgensen [9], studied by El Hélou in the case of a torus [4] etc.

Let us be more specific. From now on we suppose that $T = \mathbf{T}^d$, the d-dimensional torus, equiped with the Haar measure λ, and that the G_n are random translates of given open sets g_n, namely

$$\left. \begin{aligned} G_n &= g_n + w_n \\ G_n(t) &= g_n(t - w_n) \end{aligned} \right\} \tag{24}$$

where the w_n are independent and are distributed according to λ, and $g_n(t) = 1_{g_n}(t)$. Let us write

$$\left. \begin{aligned} v_n &= \lambda(g_n) \\ \xi_n(t) &= \int_{\mathbf{T}^d} g_n(t + s) g_n(s) ds \end{aligned} \right\}. \tag{25}$$

Then $k_N(t, s) = k_N(t - s)$ and

$$k_N(t) = \sum_{n=1}^N \left(1 + (1 - v_n)^{-2}(\xi_n(t) - v_n^2) \right). \tag{26}$$

Let us rewrite (22) in the form

$$\int \int k_N(t - s) d\sigma(t) d\sigma(s) = O(1). \tag{27}$$

Assuming

$$\lim_{n \to \infty} \text{diameter } g_n = 0 \tag{28}$$

it can be checked that (27) can be written as

$$\int \int k(t - s) d\sigma(t) d\sigma(s) < \infty \tag{29}$$

with either

$$k(t) = \prod_1^\infty (1 + \xi_n(t)) \tag{30}$$

or

$$k(t) = \exp \sum_1^\infty \xi_n(t). \tag{31}$$

(The key fact is that (27) implies $\sum v_n^2 < \infty$; this can be seen on writing

$$k_N(t) \geq \sum_{n,m \neq n} (1 - v_m)^{-2}(1 - v_n)^{-2}(\xi_m(t) - v_m^2)(\xi_n(t) - v_n^2)$$

$$\int \int (\xi_m(t - s) - v_m^2)(\xi_n(t - s) - v_n^2)d\sigma(t)d\sigma(s) =$$

$$\sum_{j \neq 0} \sum_{k \neq 0} |\hat{g}_m(j)|^2 |\hat{g}_n(k)|^2 |\hat{\sigma}(j + k)|^2$$

$$\geq |\hat{g}_m(j_0)|^2 |\hat{g}_n(-j_0)|^2 |\hat{\sigma}(0)|^2$$

$$\geq (|\hat{\sigma}(0)|^2 + o(1))v_m^2 v_n^2 \qquad (n, m \to \infty \; ; \; j_0 \text{ fixed}). \tag{32}$$

Also (29) together with either (30) or (31) implies $\sum v_n^2 < \infty$. It is easy to check that, assuming $\sum v_n^2 < \infty$, (27), (29)-(30) and (29)-(31) are equivalent. The assumption (28) is used in the last line of (32), and it can be relaxed but not removed).

From now on we assume that $k(t)$ is given by (31) or by (30), and that condition (28) is satisfied. If $\sigma = \lambda$ (the Haar measure), (29) means $k(\cdot) \in L^1(\mathbf{T}^d)$. In the general case (29) expresses that σ has finite energy with respect to the kernel $k(\cdot)$.

Therefore, if $k(\cdot) \in L^1(\mathbf{T}^d)$, $\mathbf{T}^d$ is not covered. If moreover $\operatorname{Cap}_k K > 0$ (meaning that K carries a measure of finite energy with respect to K), K is not covered.

Strikingly enough, these conditions happen to be necessary and sufficient for non-covering in the following specific cases.

1) $d = 1$, $g_n = (0, \ell_n)$. The covering of $\mathbf{T}$ is the Dvoretzky problem. (31) reads

$$k(t) = \exp \sum_1^\infty (\ell_n - |t|)^+ \quad \text{on } [-\frac{1}{2}, \frac{1}{2}], \tag{33}$$

assuming as we can $\ell_n < \frac{1}{2}$. Integrability of $k(\cdot)$ can be expressed as

$$\sum_{n=1}^\infty \frac{1}{n^2} \exp(\ell_1 + \ell_2 + \cdots + \ell_n) < \infty \tag{34}$$

assuming $\ell_1 \geq \ell_2 \geq \ell_3 \geq \cdots$

T *is covered if and only if*

$$\sum_{n=1}^{\infty} \frac{1}{n^2} \exp(\ell_1 + \ell_2 + \cdots + \ell_n) = \infty \tag{35}$$

(L. Shepp, 1972 [27]). *K is covered if and only if*

$$\mathrm{Cap}_k K = 0 \tag{36}$$

(J.-P. Kahane, 1987 [13]).

2) $d \geq 1$, g_n convex. (31) reads

$$k(t) = \exp \sum_{1}^{\infty} g_n * \tilde{g}_n(t).$$

If

$$\sum_{n=1}^{\infty} v_n^2 \exp(v_1 + v_2 + \cdots + v_n) < \infty \tag{37}$$

(Billard's condition, cf. [10]), then $k(\cdot) \in L^1(\mathbf{T}^d)$, but it is no longer a necessary and sufficient condition. In case $d = 1$, (37) implies (34) and the converse holds when $v_n = O(\frac{1}{n})$.

Now we suppose that *the g_n are homothetic simplexes. Then $\mathbf{T}^d$ is covered if and only if*

$$\int_0^1 \exp \sum_{1}^{\infty} v_n \left(1 - (\frac{1}{v_n})^{1/d}\right)^+ ds = \infty. \tag{38}$$

(38) is just another way to write $k(\cdot) \notin L^1(\mathbf{T}^d)$. The condition depends not only on the v_n's, but also on d. If it is satisfied for (v_n, d) $(d \geq 2)$, it is satisfied for $(v_n, d - 1)$. Conversely, given d, one can construct the v_n's in such a way that (38) holds for $(v_n, d - 1)$ and does not hold for (v_n, d). In other words, given the v_n, covering is easier in smaller dimensions. However all conditions (38) are equivalent (therefore equivalent with Shepp's condition

$$\sum_{1}^{\infty} \frac{1}{n^2} \exp(v_1 + v_2 + \cdots + v_n) = \infty)$$

when $v_1 \geq v_2 \geq \cdots$ and $v_n = O(\frac{1}{n})$.

Until now no example is known of a sequence g_n with diameters tending to zero for which $\mathbf{T}^d$ is not covered and $k(\cdot) \notin L^1(\mathbf{T}^d)$.

Condition (28) (diameters tending to zero) can be relaxed but cannot be removed. For, choosing a fixed g and

$$g_n(t) = g(\lambda_n t)$$

for an exponentially increasing sequence of integers λ_n, (27) can be violated and T never covered by $\bigcup(g_n + \omega_n)$ (in other words $\bigcap(f_n + \omega_n)$ never empty). Moreover, choosing a sequence λ_n very sparse, (27) can hold and (29)(30) be violated.

Proofs and comments are given in [16].

As a conclusion let us observe that we have a simple characterization of Q-regular and Q-singular measure when Q is the random operator which corresponds to case 1) $(T = \mathbf{T}$, $F_n(t) - 1 - g_n(t - \omega_n)$, $g_n(t) = 1_{(0,\ell_n)}(t)$, $k(t)$ given by (33)).

Let us recall that a measure $\sigma \in M^+(\mathbf{T})$ is k-regular if it can be written as $\sum_1^\infty \sigma_n$ and each σ_n has finite energy with respect to $k(\cdot)$. Let us say that σ is k-singular if it is carried by a Borel set of vanishing k-capacity. Then

$$Q-\text{regular} = k-\text{regular}$$

$$Q-\text{singular} = k-\text{singular} \qquad [13].$$

References

[1] P. BILLARD *Séries de Fourier aléatoirement bornées, continues, uniformément convergentes.* Ann. Scient. Ec. Norm. Sup. 82 (1965), 131-179.

[2] G. BROWN & W. MORAN *On orthogonality of Riesz products.* Proc. Cambridge Philos. Soc. 76 (1974), 173-181.

[3] A. DVORETZKY *On covering a circle by randomly placed arcs.* Proc. Nat. Acad. Sc. USA 42 (1956), 199-203.

[4] Y. EL HELOU *Recouvrement du tore T^q par des ouverts aléatoires et dimension de l'ensemble non recouvert.* C. R. Acad. Sc. Paris 287 A (1978), 815-818.

[5] FAN Ai-Hua *Une condition suffisante d'existence du moment d'ordre m (entier) du chaos multiplicatif.* Ann. Sc. Math. Québec 10 (1986), 119-120.

[6] FAN Ai-Hua Thèse, Orsay 1989.

[7] FAN Ai-Hua *Chaos additif et chaos multiplicatif de Lévy.* C. R. Acad. Sc. Paris 308 (1989), 151-154.

[8] FAN Ai-Hua *Sur la convergence de séries trigonométriques lacunaires presque partout par rapport à des produits de Riesz.* C. R. Acad. Sc. Paris 309 (1989), 295-298.

[9] J. HOFFMAN-JORGENSEN *Coverings of metric spaces by randomly placed balls.* Math. Scand. 32 (1973), 169-186.

[10] J.-P. KAHANE *Some random series of functions.* Cambridge Univ. Press 1985.

[11] J.-P. KAHANE *Sur le chaos multiplicatif.* Ann. Sc. Math. Québec 9 (1985), 105-150 and 10 (1986), 117-118.

[12] J.-P. KAHANE *Positive martingales and random measures.* Chin. Ann. Math. 8 B1 (1987), 1-12.

[13] J.-P. KAHANE *Intervalles aléatoires et décomposition des mesures.* C. R. Acad. Sc. Paris 304 (1987), 551-554.

[14] J.-P. KAHANE *Produits de poids aléatoires indépendants et applications.* Cours à l'Univ. Montréal, prépubl. Orsay 89-33.

[15] J.-P. KAHANE *Recouvrement par des simplexes homothétiques aléatoires.* C. R. Acad. Sc. Paris 310 (1990), 419-423.

[16] J.-P. KAHANE *Recouvrements aléatoires et théorie du potentiel.* Coll. Math. 50 (1990), 1-25.

[17] J.-P. KAHANE & J. PEYRIERE *Sur certaines martingales de Benoît Mandelbrot.* Advances in Math. 22 (1976), 131-145.

[18] S. J. KILMER & S. SAEKI *On Riesz product measures ; mutual absolute continuity and singularity.* Ann. Inst. Fourier, Grenoble, 38 (1988), 63-93.

[19] B. B. MANDELBROT *Possible refinement of the log-normal hypothesis concerning the distribution of energy dissipation in intermittent turbulence.* Lecture Notes in Physics, Springer Verlag 1972, 333-351.

[20] B. B. MANDELBROT *Multiplications aléatoires itérées et distributions invariantes par moyenne pondérée aléatoire.* C. R. Acad. Sc. Paris 278 (1974), 289-292 et 355-358.

[21] F. PARREAU *Ergodicité et pureté des produits de Riesz.* Ann. Inst. Fourier 40 (1990), 391-405.

[22] J. PEYRIERE *Sur les produits de Riesz.* C. R. Acad. Sc. Paris A-B 276 (1973), 1417-1419.

[23] J. PEYRIERE *Etude de quelques propriétés des produits de Riesz.* Ann. Inst. Fourier (Grenoble), 25 (1975), 127-169.

[24] J. PEYRIERE *Almost everywhere convergence of lacunary trigonometric series with respect to Riesz products.* Australian J. Math. (to appear).

[25] F. RIESZ *Uber die Fourierkoeffizienten einer stetigen Funktion von beschränkter Schwankung.* M. Zeitschrift 2 (1918), 312-315.

[20] R. SALEM *On some singular monotonic functions which are strictly increasing.* Trans. Amer. Math. Soc. 53 (1943), 427-439.

[27] L. SHEPP *Covering the circle with random arcs.* Israël J. Math. 11 (1972), 328-345.

A Method of Reduction in Harmonic Analysis
on Real Rank 1 Semisimple Lie Groups I

Takeshi KAWAZOE

Department of Mathematics, Faculty of Science and Technology,
Keio University

1. Introduction

Harmonic analysis on semisimple Lie groups G is deeply related to the unitary dual $\hat{G}$ of G, in particular the Fourier transformation of $L^2(G)$ of square-integrable functions on G is defined by using the principal series and the discrete series of G. Then, the characterization of Fourier transforms of functions in $L^2(G)$ and $C_c^\infty(G)$ of C^∞ compactly supported functions on G is one of main problems in harmonic analysis on G, so the Plancherel formula and the Paley-Wiener theorem have been studied by various people. Let K be a maximal compact subgroup of G. Since right K-invariant functions on G are regarded as functions on the symmetric space $X = G/K$, harmonic analysis on G can be regarded as one for right K-invariant functions on G. Especially, for right K-invariant functions on G the Plancherel formula consists of only wave packets defined by the principal series and thus, the discrete series does not contribute to harmonic analysis on X. This fact means that the Paley-Wiener theorem on X is simpler than the one on G, actually, it can be obtained by restricting the corresponding theorem on G to right K-invariant functions on G.

The problem treated in this paper is the reverse: is it possible to deduce the Paley-Wiener theorem for $C_c^\infty(G)$ from the one for $C_c^\infty(X)$? We would like to reduce the problem for $C_c^\infty(G,\tau)$ of functions in $C_c^\infty(G)$ with right K-type τ to the one for $C_c^\infty(G,1) = C_c^\infty(X)$ in which the discrete series does not appear. As a result, we could simplify the residue calculus in the proof of the surjectivity of the Fourier transform that is a real obstacle in the past proofs of the Paley-Wiener theorem (see [A2], [C] and [K1]). More precisely, since we know that these residues come from singularities of meromorphic functions $\Phi(\nu, a)$ and $c(-\nu)^{-1}$, which appear in the Harish-Chandra expansion of generalized τ-spherical functions on G (cf. [Wa], Chap. 9), we would like to avoid using the expansion. So far, our reduction method can be applicable to a real rank one semisimple Lie group G and a subspace $C_{c,\tau,\sigma}^\infty(G)$ of $C_c^\infty(G,\tau)$ whose principal part consists of special wave packets corresponding to the principal series $\pi(\sigma, \nu)$. Then, the Szegö operators $S_\tau(\sigma, \nu)$ will play an important role instead of the Harish-Chandra expansion. Especially, the residues which appear during the contour change come from singularities of a simple rational function $Q(\eta; \sigma, \nu)(\eta \in \hat{K}$ and see (39)) and the Plancherel measure $\mu(\sigma, \nu)$, so we can simplify the residue calculus in the proof of the Paley-Wiener theorem.

In this paper we shall give a reducion formula for $C_{c,\tau,\sigma}^\infty(G)$ (see Theorem 5.5). A proof of the Paley-Wiener theorem on G as an application of the reduction formula will appear later (see [K4]).

2. Preliminaries

2.1. Let G be a connected real rank 1 semisimple Lie group with finite center and suppose that G has a compact Cartan subgroup of G. Let $G=ANK$ be an Iwasawa decomposition of G and for $g \in G$ let $g=e^{H(g)}n(g)\kappa(g) \in ANK$ be the corresponding decomposition of g. Let M be the centralizer of A in K, $N_K(A)$ the normalizer of A in K and $W=N_K(A)/M$ the Weyl group for (G,A). Let $\mathfrak{g}=\mathfrak{a}+\mathfrak{n}+\mathfrak{k}$ be the corresponding decomposition of the Lie algebra $\mathfrak{g}$ of G and $\mathfrak{m}$ the Lie algebra of M. We denote the complexification of the Lie algebra $\mathfrak{b}$ by $\mathfrak{b}_c$ and the dual of $\mathfrak{b}$ by $\mathfrak{b}^*$. For simplicity, we put $\mathcal{F}=\mathfrak{a}^*$ and $\mathcal{F}_c=\mathfrak{a}_c^*$. Let $\mathfrak{t}_c$ and $\mathfrak{h}_c$ be the compact and noncompact Cartan subalgebras of $\mathfrak{g}_c$ respectively and $\mathfrak{h}_c^-$ the Cartan subalgebra of $\mathfrak{m}_c$ contained in $\mathfrak{h}_c$. Let β be a noncompact root for $(\mathfrak{g}_c, \mathfrak{t}_c)$ and $E_{\pm\beta}$ the root vectors satisfying $B(E_\beta, E_{-\beta}) = 2 < \beta, \beta >^1$, where $B(\cdot,\cdot)$ is the Killing form. Then $\mathfrak{a}=\mathbf{R}(E_\beta + E_{-\beta})$. We fix compatible orderings in the root spaces for $(\mathfrak{g}_c, \mathfrak{t}_c)$, $(\mathfrak{g}_c, \mathfrak{h}_c)$ and $(\mathfrak{g}_c, \mathfrak{a}_c)$ as in [KW]. Let α be the positive real simple root for $(\mathfrak{g}_c, \mathfrak{h}_c)$ and H_α the element in $\mathfrak{a}$ such that $\alpha(H_\alpha) = 2$.

Let dg denote a Haar measure on G and $L^2(G)$ the space of square-integrable functions on G with the obvious norm $\|\cdot\|_2$. Let $L^2(G)^P$ be the subspace of $L^2(G)$ consisting of wave packets and $L^2(G)^\circ$ the one consisting of cusp forms, that is, the L^2-span of matrix coefficients of the discrete series of G. Then $L^2(G)$ has a direct sum decomposition:

$$L^2(G) = L^2(G)^P \oplus L^2(G)^\circ \tag{1}$$

(see [HC3]). Let $\hat{K}$ be the set of equivalence classes of irreducible unitary representations of K. For each representative (τ, V_τ) in a class in $\hat{K}$ we denote by d_τ the dimension of the representation space V_τ and by χ_τ the character of τ. Then, for $f \in L^2(G)$ we define

$$E_\tau f(g) = d_\tau \int_K \chi_\tau(k)f(k^{-1}g)dk \quad \text{and} \quad F_\tau f(g) = d_\tau \int_K \chi_\tau(k)f(gk^{-1})dk, \tag{2}$$

where dk denotes the normalized Haar measure on K. Let $L_\tau^2(G)$ (resp. $L^2(G)_\tau$) denote the subspace of $L^2(G)$ consisting of all f in $L^2(G)$ such that $E_\tau f = f$ (resp. $F_\tau f = f$). Then $L^2(G)$ has a direct sum decomposition:

$$L^2(G) = \oplus_{\tau \in \hat{K}} L_\tau^2(G) = \oplus_{\tau \in \hat{K}} L^2(G)_\tau = \oplus_{\tau, \tau' \in \hat{K}} L_\tau^2(G)_{\tau'}, \tag{3}$$

which is called the K-type decomposition. Moreover, $L^2(G)^P$ and $L^2(G)^\circ$ (see (1)) have the same K-type decompositions.

2.2. We shall define the Szegö operator $S^\tau(\sigma, \nu)$ and give a relation between the operator and the principal series representation $\pi(\sigma, \nu)$ of G. For $(\tau, V_\tau) \in \hat{K}$ and $(\sigma, H_\sigma) \in \hat{M}$ we define $(F = C^\infty, C_c^\infty$ and $L^2)$-vector bundles on K and G by respectively

$$F(K, \sigma) = \{f \in F(K, H_\sigma); f(mk) = \sigma(m)f(k) \quad \text{for} \quad m \in M, k \in K\} \tag{4}$$

and

$$F(G,\tau) = \{f \in F(G,V_\tau); f(kg) = \tau(k)f(g) \quad \text{for} \quad k \in K, g \in G\}. \tag{5}$$

We now suppose that $[\tau,\sigma] \neq 0$, where $[\tau,\sigma]$ is the multiplicity of σ in the restriction $\tau|_M$ of τ to M, and regard H_σ as a subspace of V_τ. Then for $\nu \in \mathcal{F}_c$ the Szegö operator $S^\tau(\sigma,\nu) : C^\infty(K,\sigma) \to C^\infty(G,\tau)$ is defined by

$$S^\tau(\sigma,\nu)f(g) = \int_K e^{(\rho+\sqrt{-1}\nu)H(kx^{-1})}\tau(\kappa(kg^{-1}))^{-1}f(k)dk \tag{6}$$

(see [KW]). On the other hand, let $(\pi(\sigma,\nu), C^\infty(K,\sigma))$ $(\sigma \in \hat{M}, \nu \in \mathcal{F}_c)$ denote the nonunitary principal series of G, which operates in $C^\infty(K,\sigma)$ with the G-action given by

$$\pi(\sigma,\nu;g)f(k) = e^{(\rho-\sqrt{-1}\nu)H(kg)}f(\kappa(kg)) \quad (g \in G). \tag{7}$$

Obviously, $\pi(\sigma,\nu)$ is unitary if and only if $\nu \in \mathcal{F}$. Then the relation between the Szegö operator $S^\tau(\sigma,\nu)$ and the principal series $\pi(\sigma,\nu)$ is given by the following,

LEMMA 2.1 ([KW], LEMMA 6.2). For $f \in C_c^\infty(G)$ and $\nu \in \mathcal{F}_c$

$$S^\tau(\sigma,\nu)f(g) = \int_K \tau(k)^{-1}\pi(\sigma,\nu;g)f(k)dk.$$

2.3. In the following arguments we need the explicit form of a complete orthonormal system of $L^2(K,\sigma)$. For $(\sigma,H_\sigma) \in \hat{M}$ we let $\hat{K}_\sigma = \{\tau \in \hat{K}; [\tau,\sigma] \neq 0\}$ and then, for $(\tau,V_\tau) \in \hat{K}_\sigma$ we denote the set of irreducible constituents of $\tau|_M$ that are equivalent with σ by $\{(\sigma(\alpha), H_{\sigma(\alpha)}); 1 \leq \alpha \leq [\tau,\sigma]\}$. Let $\{v_{\tau,i}; 1 \leq i \leq d_\tau\}$ and $\{h_{\sigma(\alpha),i}; 1 \leq i \leq d_\sigma\}$ denote orthonormal basis of V_τ and $H_{\sigma(\alpha)}$ respectively. Then, since we can regard $H_{\sigma(\alpha)}$ as a subspace of V_τ, without loss of generality we may assume that

$$v_{\tau,(\alpha-1)d_\sigma+i} = h_{\sigma(\alpha),i} \quad (1 \leq \alpha \leq [\tau,\sigma], 1 \leq i \leq d_\sigma). \tag{8}$$

For simplicity, we denote the set of integers $\{1,2,...,N\}$ by I_N and put $I_\sigma = I_{d_\sigma}$ and $I_\tau = I_{d_\tau}$. Then for $i \in I_\tau$ and $\beta \in I_{[\tau,\sigma]}$ we define

$$\phi_{\tau,i,\beta}(k) = d_\tau^{1/2}d_\sigma^{-1/2}\sum_{j \in I_\sigma} <\tau(k)v_{\tau,i}, h_{\sigma(\beta),j} > h_{\sigma(\beta),j} \quad (k \in K). \tag{9}$$

Since $H_{\sigma(\beta)}$ can be identified with H_σ, the functions $\phi_{\tau,i,\beta}$ belong to $L^2(K,\sigma)$ and form a complete orthonormal system of $L^2(K,\sigma)$. It follows from Lemma 2.1 that

$$S^\tau(\sigma(\beta),\nu)f(g) = \int_K \tau(k)^{-1}\pi(\sigma,\nu;g)f(k)dk$$

$$= \sum_{s \in I_\tau} <\pi(\sigma,\nu;g)f(k), P_\beta(\tau(k)v_{\tau,s}) > dkv_{\tau,s} \tag{10}$$

$$= \sum_{s \in I_\tau} <\pi(\sigma,\nu;g)f, \phi_{\tau,s,\beta} > v_{\tau,s},$$

where $< \cdot , \cdot >$ denotes the inner product of $L^2(K, \sigma)$ and P_β the orthogonal projection of V_τ onto $H_{\sigma(\beta)}$. When $\sigma = 1$, it is well known that $[\eta, 1] = 1$ for $\eta \in \hat{K}_1$ (see [Ko]), so we denote the complete orthonormal system of $L^2(K, 1)$ obtained above by $\{\psi_{\eta,p}; p \in I_\eta, \eta \in \hat{K}_1\}$.

2.4. In order to put a reduction formula into action we assume the following results: (i) harmonic analysis on the symmetric space $X = K\backslash G$, (ii) the fact that L^2-functions f on G can be written as the sum of wave packets f^P and cusp forms f° (see (1)) and (iii) the Fourier analysis of f^P. Here we shall summarize some results about the Fourier analysis of f^P and describe the Fourier inversion formula by using the Szegö operator defined in 2.2.

Let $\sigma \in \hat{M}$ and $\nu \in \mathcal{F}_c$. Then for $f \in C_c^\infty(G)$ we define

$$\hat{f}_{\tau,i,\alpha;\tau',i',\alpha'}(\sigma, \nu) = \int_G f(g) < \pi(\sigma, \nu; g^{-1})\phi_{\tau,i,\alpha}, \phi_{\tau',i',\alpha'} > dg, \qquad (11)$$

where $\tau, \tau' \in \hat{K}_\sigma, i \in I_\tau, i' \in I_{\tau'}, \alpha \in I_{[\tau,\sigma]}$ and $\alpha' \in I_{[\tau',\sigma]}$ and moreover,

$$\hat{f}(\phi; \sigma, \nu)(k) = \int_G f(g)\pi(\sigma, \nu; g^{-1})\phi(k)dg \quad (k \in K) \qquad (12)$$

for $\phi \in L^2(K, \sigma)$. As a function of ν, $\hat{f}_{\tau,i,\alpha;\tau',i',\alpha'}(\sigma, \nu)$ is a holomorphic function of exponential type and $\hat{f}(\phi; \sigma, \nu)(k)$ is one of uniform exponential type in the sense of [H4], p.133. In particular, for $\phi = \phi_{\tau,i,\alpha}$

$$\hat{f}(\phi_{\tau,i,\alpha}; \sigma, \nu)(k) = \sum_{\tau',i',\alpha'} \hat{f}_{\tau,i,\alpha;\tau',i',\alpha'}(\sigma, \nu)\phi_{\tau',i',\alpha'}(k), \qquad (13)$$

where the sum is taken over all $\tau' \in \hat{K}_\sigma$, $i' \in I_{\tau'}$ and $\alpha' \in I_{[\tau',\sigma]}$. We now write down the Fourier inversion formula for f^P by using the Szegö operator $S^\tau(\sigma, \nu)$ (see (6)). Applying (10) and (13), we can deduce that

$$S^\tau(\sigma(\alpha), \nu)(\hat{f}(\phi_{\tau,i,\alpha}; \sigma, \nu))(g)$$
$$= \sum_{\tau',i',\alpha'} \hat{f}_{\tau,i,\alpha;\tau',i',\alpha'}(\sigma, \nu) \sum_{s \in I_\tau} < \pi(\sigma, \nu; g)\phi_{\tau',i',\alpha'}, \phi_{\tau,s,\alpha} > v_{\tau,s}. \quad (14)$$

As a V_τ-valued function of ν, this function is holomorphic of exponential type. Therefore, according to the inversion formula given in [A1], §3, we can rewrite the principal part f^P of f in $C_c^\infty(G)$ as follows.

LEMMA 2.2 ([A1], **LEMMA 5**). *There exists a nonnegative function $\mu(\sigma, \nu)$ on $\hat{M} \times \mathcal{F}$ such that for $f \in C_c^\infty(G)$*

$$f^P(g) = \sum_{\sigma \in \hat{M}} \sum_{\tau,i,\alpha} \sum_{\tau',i',\alpha'} \int_{\mathcal{F}} \hat{f}_{\tau,i,\alpha;\tau',i',\alpha'}(\sigma, \nu) < \pi(\sigma, \nu; g)\phi_{\tau',i',\alpha'}, \phi_{\tau,i,\alpha} > \mu(\sigma, \nu)d\nu$$
$$= \sum_{\sigma \in \hat{M}} \sum_{\tau,i,\alpha} \int_{\mathcal{F}} < S^\tau(\sigma(\alpha), \nu)(\hat{f}(\phi_{\tau,i,\alpha}; \sigma, \nu))(g), v_{\tau,i} > \mu(\sigma, \nu)d\nu.$$

This μ-function is explicitly obtained by [A1], §3 and [HC3], §36. Furthermore, the Plancherel formula for $L^2(G)^P$ can be stated as

LEMMA 2.3 ([A1], THEOREM 2). *For $f \in C_c^\infty(G)$*

$$\int_G |f^P(g)|^2 dg = \sum_{\sigma \in \hat{M}} \sum_{\tau,i,\alpha} \sum_{\tau',i',\alpha'} \int_{\mathcal{F}} |\hat{f}_{\tau,i,\alpha;\tau',i',\alpha'}(\sigma,\nu)|^2 \mu(\sigma,\nu)d\nu$$

$$= \sum_{\sigma \in \hat{M}} \sum_{\tau,i,\alpha} \int_{\mathcal{F}} \int_K \|\hat{f}(\phi_{\tau,i,\alpha};\sigma,\nu)(k)\|^2 \mu(\sigma,\nu)dkd\nu,$$

where $\| \cdot \|$ is the norm in H_σ.

Last we note the W-invariance of the Fourier transform. Let $w \in W$ be a nontrivial representative of $W = N_K(A)/M$ such that $w\sigma = \sigma$ (see [Kn], Lemma 4). Then for each $\nu \in \mathcal{F}$ the unitary principal series $\pi(\sigma,\nu)$ and $\pi(\sigma,w\nu)$ are unitary equivalent, that is, there exists a unitary intertwining operator $A(w,\sigma,\nu)$ of $L^2(K,\sigma)$ such that $A(w,\sigma,\nu)\pi(\sigma,\nu) = \pi(\sigma,w\nu)A(w,\sigma,\nu)$ (see [KS]), especially, it does not change the K-type τ of $\phi_{\tau,i,\alpha}$. Therefore, applying the unitary intertwining operator to the right hand side of

$$\sum_{i \in I_\tau, \alpha \in I_{[\tau,\sigma]}} < S^\tau(\sigma(\alpha),\nu)(\hat{f}(\phi_{\tau,i,\alpha};\sigma,\nu))(g), v_{\tau,i} > \tag{15}$$

$$= \sum_{i \in I_\tau, \alpha \in I_{[\tau,\sigma]}} \sum_{\tau',i',\alpha'} \int_G f(g) < \pi(\sigma,\nu;g^{-1})\phi_{\tau,i,\alpha}, \phi_{\tau',i',\alpha'} > dg$$

$$\times < \pi(\sigma,\nu;g)\phi_{\tau',i',\alpha'}, \phi_{\tau,i,\alpha} >$$

(see (11) and (14)), we can deduce the W-invariance of the first line (15) such as

LEMMA 2.4. *For $f \in C_c^\infty(G)$ and $\nu \in \mathcal{F}_c$*

$$\sum_{i \in I_\tau, \alpha \in I_{[\tau,\sigma]}} < S^\tau(\sigma(\alpha),w\nu)(\hat{f}(\phi_{\tau,i,\alpha};\sigma,w\nu))(g), v_{\tau,i} >$$

$$= \sum_{i \in I_\tau, \alpha \in I_{[\tau,\sigma]}} < S^\tau(\sigma(\alpha),\nu)(\hat{f}(\phi_{\tau,i,\alpha};\sigma,\nu))(g), v_{\tau,i} > .$$

2.5 As in the case of the Euclidean space, all equations of $\hat{f}(\phi;\sigma,\nu)(\nu \in \mathcal{F})$ for $f \in C_c^\infty(G)$ can be extended to the ones for $f \in L^2(G)$ by the usual manner. Therefore, for $f \in L^2(G)$ we can define

$$f^P_{\tau,\sigma(\alpha)}(g) = \sum_{i \in I_\tau} \int_{\mathcal{F}} < S^\tau(\sigma(\alpha),\nu)(\hat{f}(\phi_{\tau,i,\alpha};\sigma,\nu))(g), v_{\tau,i} > \mu(\sigma,\nu)d\nu \tag{16}$$

and thus, Lemma 2.2 implies that

$$f^P = \sum_{\sigma \in \hat{M}, \tau \in \hat{K}_\sigma, \alpha \in I_{[\tau,\sigma]}} f^P_{\tau,\sigma(\alpha)}. \tag{17}$$

We now define $L^2_{\tau,\sigma(\alpha)}(G)^P$ by the subspace of $L^2_\tau(G)^P$ (see 2.1) consisting of all wave packets $f^P_{\tau,\sigma(\alpha)}$ of f in $L^2(G)^P$. Then (17) implies that $L^2(G)^P$ has a direct sum decomposition:

$$L^2(G)^P = \oplus_{\sigma \in \dot{M}, \tau \in \dot{K}_\sigma, \alpha \in I_{[\tau,\sigma]}} L^2_{\tau,\sigma(\alpha)}(G)^P. \tag{18}$$

On the other hand, we define

$$C^\infty_{c,\tau,\sigma(\alpha)}(G) = \{f \in C^\infty_c(G); f^P \in L^2_{\tau,\sigma(\alpha)}(G)^P\} \tag{19}$$

and $L^2_{\tau,\sigma(\alpha)}(G)^\circ$ the L^2-completion of $C^\infty_{c,\tau,\sigma(\alpha)}(G)^\circ = \{f^\circ; f \in C^\infty_{c,\tau,\sigma(\alpha)}(G)\}$.

LEMMA 2.5. *The discrete part f° of $f \in C^\infty_c(G)$ is uniquely determined by the principal part f^P of f.*

Proof. For $f \in C^\infty_c(G)$ we denote the K-type decomposition of f as $f = \sum_{\tau,\tau' \in \dot{K}} f_{\tau,\tau'}$, where $f_{\tau,\tau'} = E_\tau F_{\tau'} f$ (see (3)). Then, we see that $f_{\tau,\tau'} = f^P_{\tau,\tau'} + f^\circ_{\tau,\tau'} \in C^\infty_c(G)$ and $f^\circ_{\tau,\tau'}$ is real analytic on G (see [HC1], §27). Therefore, each $f^\circ_{\tau,\tau'}$ is uniquely determined by $f^P_{\tau,\tau'}$ and thus, the desired result follows. ∎

According to Lemma 2.5, (18) and (19), we see that $L^2(G)^\circ$ also has the following direct sum decomposition:

$$L^2(G)^\circ = \oplus_{\sigma \in \dot{M}, \tau \in \dot{K}_\sigma, \alpha \in I_{[\tau,\sigma]}} L^2_{\tau,\sigma(\alpha)}(G)^\circ. \tag{20}$$

Then we put

$$L^2_{\tau,\sigma(\alpha)}(G) = L^2_{\tau,\sigma(\alpha)}(G)^P \oplus L^2_{\tau,\sigma(\alpha)}(G)^\circ. \tag{21}$$

When $\sigma = 1$, it is well known that $[\eta, 1] = 1$ for $\eta \in \hat{K}_1$, so, for simplicity we shall denote $L^2_{\eta,1(1)}(G)$ and $C^\infty_{c,\eta,1(1)}(G)$ by $L^2_{\eta,1}(G)$ and $C^\infty_{c,\eta,1}(G)$ respectively. Especially, when $\sigma = 1$ and $\eta = 1$, we use the notations $L^2(K\backslash G)$ and $C^\infty_c(K\backslash G)$ instead of $L^2_{1,1}(G)$ and $C^\infty_{c,1,1}(G)$ respectively.

3. Reduction formula for Szegö operators

We fix $\sigma \in \hat{M}$ and $\tau \in \hat{M}_\sigma$, and regard H_σ as a subspace of V_τ. Let $\lambda_\tau \in \mathfrak{t}^*_c$ denote the highest weight of τ and λ_σ a weight of τ that is highest in H_σ. For each weight γ we denote by ϕ_γ a corresponding normalized weight vector. In the rest of the paper we assume that a finite dimensional $\mathfrak{g}$-module π lifts to a representation of G by passing to a finite covering of G and moreover,

(A) σ is the restriction of τ to the M-cyclic subsapce generated by ϕ_{λ_τ}, that is, $\lambda_\tau = \lambda_\sigma$.

LEMMA 3.1. *There exist a finite dimensional representation (π, U_π) of G in which $\sigma \in \hat{M}$ and $\tau \in \hat{M}_\sigma$ are embedded and there exists a weight $-\mu \in \mathfrak{h}_c^*$ of π satisfying*

(i) $-\mu$ *is lowest with respect to* $(\mathfrak{g}_c, \mathfrak{a}_c)$,

(ii) $\lambda_\sigma |_{\mathfrak{h}_c^-} = -\mu |_{\mathfrak{h}_c^-}$,

(iii) $P_\tau(\phi_{-\mu}) = c\phi_{\lambda_\sigma}$ $(c \neq 0)$,

where $P_\tau : U_\pi \to V_\tau$ is the orthogonal projection of U_π to V_τ.

Proof. Let π be a finite dimensional representation of G with the highest weight $\lambda = \lambda_\tau = \lambda_\sigma \in \mathfrak{t}_c^*$ and $Ad(\gamma) : \mathfrak{t}_c \to \mathfrak{h}_c$ the Cayley transformation (cf. [KW], p.176 and [HC], §30). Then, noting the compatible orderings in the root spaces for $(\mathfrak{g}_c, \mathfrak{t}_c)$ and $(\mathfrak{g}_c, \mathfrak{a}_c)$, we see that $\pi(\gamma)\phi_\lambda$ is a lowest weight vector with respect to $(\mathfrak{g}_c, \mathfrak{a}_c)$, so we put $\phi_{-\mu} = \pi(\gamma)\phi_\lambda = \phi_{Ad(\gamma)\lambda}$. Since $Ad(\gamma)$ is trivial on $\mathfrak{h}_c^-$ and $\gamma = exp\pi(E_\beta - E_{-\beta})/4$ (cf. [KW], p.176), it follows that $-\mu |_{\mathfrak{h}_c^-} = Ad(\gamma)\lambda |_{\mathfrak{h}_c^-} = \lambda |_{\mathfrak{h}_c^-}$ and $P_\tau(\phi_{-\mu}) = P_\tau(\pi(\gamma)\phi_\lambda) = c\phi_\lambda$ $(c \neq 0)$. Therefore, π and $-\mu$ satisfy the desired conditions. ∎

Remark 3.2. When $G = SU(n, 1)$ $(n \in \mathbf{N})$, $K = S(U(n) \times U(1))$ is not semisimple and thus, it has one dimensional representations τ_ℓ $(\ell \in \mathbf{Z})$ defined by

$$\tau_\ell \quad : \quad \begin{pmatrix} U & 0 \\ 0 & |U|^{-1} \end{pmatrix} \quad \longmapsto \quad |U|^{-\ell}$$

for $U \in U(n)$. By choosing a suitable ordering for $(\mathfrak{g}_c, \mathfrak{t}_c)$, without loss of generality, we may assume that $\ell \geq 0$. Then, G acts naturally on the space U_ℓ of all homogeneous polynomials of degree ℓ in the variables $z_1, z_2, \ldots z_{n+1}$, so we get a finite dimensional representation (π_ℓ, U_ℓ) of G. By regarding V_{τ_ℓ} and H_{σ_ℓ} $(\sigma_\ell = \tau_\ell |_M)$ as $\mathbf{C}(z_{n+1})^\ell$, τ_ℓ and σ_ℓ can be embedded in π_ℓ. We now take $\mathfrak{a} = \mathbf{R}(e_1 \otimes e_{n+1}^* + e_{n+1} \otimes e_1^*)$ (see [JW]) and put $\phi_0 = (z_1 + z_{n+1})^\ell \in U_\ell$. Then, ϕ_0 is a lowest weight vector with the lowest weight $-\ell\alpha$ with respect to $(\mathfrak{g}_c, \mathfrak{a}_c)$, $P_\tau(\phi_0) = (z_{n+1})^\ell$ and $\pi_\ell(m)\phi_0 = \sigma_\ell(m)$ $(m \in M)$. Therefore, π_ℓ, ϕ_0 and $\mu = \ell\alpha$ satisfy the desired properties stated in Lemma 3.1. This means that all arguments below are also applicable to $\sigma_\ell \in \hat{M}$ and $\tau_\ell \in \hat{K}_{\sigma_\ell}$.

By Lemma 3.1 we can regard H_σ and V_τ as the subspaces of U_π and σ as $\sigma(1)$ in the previous notation. For simplicity, we shall denote $\sigma(1)$ as σ. Especially, $\{v_{\tau,i}; i \in I_\tau\}$ and $\{v_{\tau,i}; i \in I_\sigma\}$ are orthonormal systems of V_τ and H_σ respectively (see (8)). Let H_{σ_0} be the M-cyclic subsapce of U_π generated by $\phi_{-\mu}$ and σ_0 the restriction of π to H_{σ_0}. Then it follows from Lemma 3.1 (ii) that (σ, H_σ) is equivalent to (σ_0, H_{σ_0}). So, we let $\{u_{\pi,i}; i \in I_\sigma\}$ be an orthonormal system of H_{σ_0} corresponding to $\{v_{\tau,i}; i \in I_\sigma\}$, and we extend it to an orthonormal system of U_π denoted as $\{u_{\pi,i}; i \in I_\pi\}$ $(I_\pi = I_{dim\pi})$. Let $\phi = \phi_{\lambda_\sigma}$ and $\phi_0 = \phi_{-\mu}$. Then, it follows from Lemma 3.1 (iii) that

LEMMA 3.3. *There exists a nonzero constant c such that*

$$< v, \phi_0 >= c < v, \phi > \quad \text{for all } v \in V_\tau.$$

In particular,

$$< v, u_{\pi,i} >= c < v, v_{\tau,i} > \quad \text{for all } i \in I_\sigma, v \in V_\tau.$$

We now investigate the relation between the Szegö operators $S^\tau(\sigma, \nu)$ and $S^1(1, \nu - \sqrt{-1}\mu_\alpha)$, where $\mu_\alpha =< \mu, \alpha > \alpha < \alpha, \alpha >^{-1}$ and 1's denote the trivial representations of K and M respectively. For $f \in C^\infty(K, \sigma)$ we define

$$\tilde{f}(k) = \sum_{i \in I_\sigma} < f(k), v_{\tau,i} > u_{\pi,i} \quad (k \in K). \tag{23}$$

Then $\tilde{f}$ is an H_{σ_0}-valued function on K and by Lemma 3.3 it satisfies

$$< v, f >= c^{-1} < v, \tilde{f} > \quad \text{for all } v \in V_\tau. \tag{24}$$

Here we note that

$$< \phi_0, \pi(g)u >= e^{-\mu_\alpha H(a)} < \phi_0, \pi(k)u > \quad (u \in U_\pi) \tag{25}$$

for $g = ank \in G = ANK$, because $\phi_0 = \phi_{-\mu}$ is a weight vector with the lowest weight $-\mu|_{a_c}$ with respect to $(\mathfrak{g}_c, \mathfrak{a}_c)$ (see Lemma 3.1 (i)). Therefore, since $\tilde{f}$ is an H_{σ_0}-valued function on K and H_{σ_0} is the M-cyclic subspace of U_π generated by ϕ_0, it follows from (24) and (25) that

LEMMA 3.4. *For each $i \in I_\tau$*

$$< \tilde{f}(k), \pi(kg^{-1})v_{\tau,i} > = e^{-\mu_\alpha H(kg^{-1})} < \tilde{f}(k), \pi(\kappa(kg^{-1}))v_{\tau,i} >$$
$$= e^{-\mu_\alpha H(kg^{-1})}c < f(k), \tau(\kappa(kg^{-1}))v_{\tau,i} > .$$

Then, substituting this equation for (6), we can deduce that

$$< S^\tau(\sigma, \nu)f(g), v_{\tau,i} >$$
$$= \int_K e^{(\rho+\sqrt{-1}\nu)H(kg^{-1})} < f(k), \tau(\kappa(kg^{-1}))v_{\tau,i} > dk$$
$$= \int_K e^{(\rho+\sqrt{-1}\nu+\mu_\alpha)H(kg^{-1})}c^{-1} < \tilde{f}(k), \pi(kg^{-1})v_{\tau,i} > dk \tag{26}$$
$$= \sum_{j \in I_\pi} \Pi_{ij}(g) \int_K e^{(\rho+\sqrt{-1}\nu+\mu_\alpha)H(kg^{-1})}c^{-1} < \pi(k)^{-1}\tilde{f}(k), u_{\pi,j} > dk$$
$$= \sum_{j \in I_\pi} \Pi_{ij}(g) \int_K e^{(\rho+\sqrt{-1}\nu+\mu_\alpha)H(kg^{-1})} \tilde{f}_j(k)dk,$$

where

$$\Pi_{ij}(g) =< u_{\pi,j}, \pi(g^{-1})v_{\tau,i} > \tag{27}$$

and

$$\tilde{f}_j(k) = c^{-1} < \pi(k)^{-1}\hat{f}(k), u_{\pi,j} > \quad (k \in K) \tag{28}$$

for $i \in I_\tau$ and $j \in I_\pi$. Then each $\tilde{f}_j$ belongs to $C^\infty(K,1)$. Actually, since σ is equivalent to σ_0, it follows that $\tilde{f}(mk) = \sum_{i \in I_\sigma} < \sigma(m)f(k), v_{\tau,i} > u_{\pi,i} = \sum_{i \in I_\sigma} < f(k), v_{\tau,i} > \sigma_0(m)u_{\pi,i} = \sigma_0(m)\tilde{f}(k)(m \in M)$ and thus, $\tilde{f}_j(mk) = < \pi(k)^{-1}\pi(m)^{-1}\sigma_0(m)\tilde{f}(k), u_{\pi,j} > = \tilde{f}_j(k)$. Therefore, the last integral in (26) is noting but $S^1(1, \nu - \sqrt{-1}\mu_\alpha)\tilde{f}_j(g)$ (see (6)) and thus, we obtain the following,

PROPOSITION 3.5. *Let* $f \in C^\infty(K,\sigma)$. *Then for each* $i \in I_\tau$ *and* $\nu \in \mathcal{F}_c$

$$< S^\tau(\sigma,\nu)f(g), v_{\tau,i} >= \sum_{j \in I_\pi} \Pi_{ij}(g)S^1(1, \nu - \sqrt{-1}\mu_\alpha)\tilde{f}_j(g).$$

4. Reduction formula for Fourier transforms

Let f be in $C_c^\infty(G)$ and $\hat{f}(\phi_{\tau,i,1}; \sigma, \nu)$ $(\sigma(1) = \sigma)$ the Fourier transform of f defined by (12). For a fixed $\nu \in \mathcal{F}_c$ this Fourier transform belongs to $C^\infty(K,\sigma)$, so we can apply Proposition 3.5 to $\hat{f}(\phi_{\tau,i,1}; \sigma, \nu)$. Especially, each $\tilde{\hat{f}}_j(\phi_{\tau,i,1}; \sigma, \nu)$ $(j \in I_\pi$ and see (28)) belongs to $C^\infty(K,1)$. In this section we shall rewrite $\tilde{\hat{f}}_j(\phi_{\tau,i,1}; \sigma, \nu)$ in terms of the Fourier transform associated with $\pi(1, \nu - \sqrt{-1}\mu_\alpha)$. Let

$$\tilde{\hat{f}}_j(\phi_{\tau,i,1}; \sigma, \nu) = \sum_{\eta,p}(\tilde{\hat{f}}_j(\phi_{\tau,i,1}; \sigma, \nu), \psi_{\eta,p})\psi_{\eta,p} \tag{29}$$

denote the Fourier series expansion of $\tilde{\hat{f}}_j(\phi_{\tau,i,1}; \sigma, \nu)$, where the sum is taken over all $\eta \in \hat{K}_1$ and $p \in I_\eta$, and $\{\psi_{\eta,p}\}$ is a complete orthonormal system of $C^\infty(K,1)$ obtained in 2.3. Then, substituting (28), (23), (12) and (7) for the avove Fourier series expansion, we can deduce that

$$\tilde{\hat{f}}_j(\phi_{\tau,i,1}; \sigma, \nu) = \sum_{\eta,p} \int_K c^{-1} < \pi(k)^{-1}\hat{f}(\phi_{\tau,i,1}; \sigma, \nu)(k), u_{\pi,j} > \overline{\psi_{\eta,p}(k)}dk\psi_{\eta,p}$$

$$= \sum_{\eta,p} \int_G f(g) \int_K e^{(\rho-\sqrt{-1}\nu)H(kg^{-1})}c^{-1} < \tilde{\phi}_{\tau,i,1}(\kappa(kg^{-1})), \pi(k)u_{\pi,j} > \overline{\psi_{\eta,p}(k)}dkdg\psi_{\eta,p}.$$

Then by the change of variable formula (cf. [KW], Lemma 6.2)

$$= \sum_{\eta,p} \int_G f(g) \int_K e^{(\rho+i\nu)H(kg)}c^{-1} < \tilde{\phi}_{\tau,i,1}(k), \pi(\kappa(kg))u_{\pi,j} > \overline{\psi_{\eta,p}(\kappa(kg))}dkdg\psi_{\eta,p}.$$

Therefore, applying Lemma 3.4 and again using the change of variable formula, we see that

$$= \sum_{\eta,p} \int_G f(g) \int_K e^{(\rho + \sqrt{-1}(\nu - \sqrt{-1}\mu_\alpha))H(kg)}$$

$$\times c^{-1} < \tilde{\phi}_{\tau,i,1}(k), \pi(kg)u_{\pi,j} > \overline{\psi_{\eta,p}(\kappa(kg))} dk dg \psi_{\eta,p}$$

$$= \sum_{\eta,p} \sum_{s \in I_\pi} \int_G f(g)\overline{\pi_{sj}(g)} \int_K e^{(\rho + \sqrt{-1}(\nu - \sqrt{-1}\mu_\alpha))H(kg)}$$

$$\times c^{-1} < \pi(k)^{-1}\tilde{\phi}_{\tau,i,1}(k), u_{\pi,s} > \overline{\psi_{\eta,p}(\kappa(kg))} dk dg \psi_{\eta,p}$$

$$= \sum_{\eta,p} \sum_{s \in I_\pi} \int_G f(g)\overline{\pi_{sj}(g)} \int_K e^{(\rho - \sqrt{-1}(\nu - \sqrt{-1}\mu_\alpha))H(kg^{-1})}$$

$$\times c^{-1} < \pi(\kappa(kg^{-1}))^{-1}\tilde{\phi}_{\tau,i,1}(\kappa(kg^{-1})), u_{\pi,s} > \overline{\psi_{\eta,p}(k)} dk dg \psi_{\eta,p},$$

where

$$\pi_{sj}(g) = < \pi(g)u_{\pi,j}, u_{\pi,s} > \tag{30}$$

for $s, j \in I_\pi$. Then, by using (28), (7) and (12) again we can reformulate the last integrals as

$$= \sum_{s \in I_\pi} \int_G f(g)\overline{\pi_{sj}(g)}\pi(1, \nu - \sqrt{-1}\mu_\alpha; g^{-1})\tilde{\phi}_{\tau,i,1,s} dg$$

$$= \sum_{s \in I_\pi} (\overline{\pi_{sj}}f)^\wedge(\tilde{\phi}_{\tau,i,1,s}; 1, \nu - \sqrt{-1}\mu_\alpha).$$

PROPOSITION 4.1. *Let $f \in C_c^\infty(G)$. Then for each $j \in I_\pi$ and $\nu \in \mathcal{F}_c$*

$$\tilde{f}_j(\phi_{\tau,i,1}; \sigma, \nu)(k) = \sum_{s \in I_\pi} (\overline{\pi_{sj}}f)^\wedge(\tilde{\phi}_{\tau,i,1,s}; 1, \nu - \sqrt{-1}\mu_\alpha)(k).$$

5. Main theorems

We retain the notations in the previous sections and suppose that $\sigma \in \hat{M}$ and $\tau \in \hat{K}_\sigma$ satisfy the assumption (A) in §3.

5.1. Combining Proposition 3.5 with Proposition 4.1, we can obtain the following reduction formula.

PROPOSITION 5.1. *Let $f \in C_c^\infty(G)$. Then for each $i \in I_\tau$ and $\nu \in \mathcal{F}_c$*

$$< S^\tau(\sigma, \nu + \sqrt{-1}\mu_\alpha)\hat{f}(\phi_{\tau,i,1}; \sigma, \nu + \sqrt{-1}\mu_\alpha)(g), v_{\tau,i} >$$

$$= \sum_{j \in I_\tau} \Pi_{ij}(g)S^1(1, \nu)\hat{\tilde{f}}_j(\phi_{\tau,i,1}; \sigma, \nu + \sqrt{-1}\mu_\alpha)(g)$$

$$= \sum_{j,s \in I_\tau} \Pi_{ij}(g)S^1(1, \nu)(\overline{\pi_{sj}}f)^\wedge(\tilde{\phi}_{\tau,i,1,s}; 1, \nu)(g).$$

Unfortunately, as a function of ν, $S^1(1, \nu)(\overline{\pi_{sj}}f)^\wedge(\tilde{\phi}_{\tau,i,1,s}; 1, \nu)$ is not of the form of a W-invariant function (see Lemma 2.4). Then, as we carry out the Fourier analysis of the right hand side, this fact is an obstacle, so we must remove it. In the following, by applying a differential operator on G to the right hand side we shall extract a W-invariant function of ν.
We put

$$\hat{K}_{1,\tau} = \{\eta \in \hat{K}_1; \quad [\pi^* \otimes \tau, \eta] \neq 0\}. \tag{31}$$

Then, it follows from the definition (28) that the Fourier series expansion of $\tilde{\phi}_{\tau,i,1,s}$ ($s \in I_\pi$) is given as

$$\tilde{\phi}_{\tau,i,1,s}(k) = \sum_{\eta \in \hat{K}_{1,\tau}, p \in I_\eta} C_{\tau,i,s;\eta,p}\psi_{\eta,p}(k) \quad (k \in K) \tag{32}$$

and then, the Szegö operator in the right hand side of Proposition 5.1 is rewritten as

$$S^1(1, \nu)(\overline{\pi_{sj}}f)^\wedge(\tilde{\phi}_{\tau,i,1,s}; 1, \nu) = \sum_{\eta \in \hat{K}_{1,\tau}, p \in I_\eta} C_{\tau,i,s;\eta,p}S^1(1, \nu)(\overline{\pi_{sj}}f)^\wedge(\psi_{\eta,p}; 1, \nu). \tag{33}$$

We now rewrite the right hand side of (33) as the Fourier transform corresponding to $\psi_{1,1}$ instead of $\psi_{\eta,p}$. We regard each element X in $U(\mathfrak{g}_c)$, the universal enveloping algebra of $\mathfrak{g}_c$, as a right invariant differential operator on G by the usual manner and denote it by the same letter. Moreover, for $\pi \in \hat{G}$ we denote by $d\pi$ the differential representation of $U(\mathfrak{g}_c)$. For each $\eta \in \hat{K}_1$ we may choose $\psi_{\eta,1}$ as the M-fixed vector in $\{\psi_{\eta,p}; p \in I_\eta\}$. Then, noting the explicit realization of $\pi(1, \nu)$ on $L^2(M \backslash K) = L^2(K, 1)$ (cf. [JW], Theorem 4.1 and [J], Theorem 4.1), we see the following,

LEMMA 5.2. *There exist a nonnegative integer $n(\eta)$ and a nonzero polynomial $P_\eta(\nu)$ ($\nu \in \mathcal{F}_c$) with zeros on the imarginary axis such that*

$$d\pi(1, \nu; H_\alpha^{n(\eta)})\psi_{1,1} = P_\eta(\nu)\psi_{\eta,1} + \quad \ll \text{other } K\text{-types} \gg$$

where $\ll$ other K-types $\gg$ is a linear combination of $\psi_{\eta',1}$ with $\eta' \neq \eta \in \hat{K}_1$.

Let E_{ij}^ξ and F_{ij}^ξ ($\xi \in \hat{K}, i, j \in I_\xi$) denote the operators on $C(G)$ defined by

$$E_{ij}^\xi h(g) = d_\xi \int_K \overline{\xi_{ij}}(k)h(kg)dk \quad \text{and} \quad F_{ij}^\xi h(g) = d_\xi \int_K \overline{\xi_{ij}}(k)h(gk)dk \tag{34}$$

for $h \in C(G)$. Then, applying Lemma 5.2 and (34), we see that for each $\eta \in \hat{K}_{1,\tau}$ and $p \in I_\eta$

$$((-H_\alpha)^{n(\eta)} E^\eta_{p1}(\overline{\pi_{sj}f}))^\wedge(\psi_{1,1}; 1, \nu)(k)$$

$$= \int_G ((-H_\alpha)^{n(\eta)} E^\eta_{p1}(\overline{\pi_{sj}f}))(g)\pi(1, \nu; g^{-1})\psi_{1,1}(k)dg \quad \text{(see (12))}$$

$$= \int_G E^\eta_{p1}(\overline{\pi_{sj}f})(g)\pi(1, \nu; g^{-1})d\pi(1, \nu; H^{n(\eta)}_\alpha)\psi_{1,1}(k)dg$$

$$= \int_G E^\eta_{p1}(\overline{\pi_{sj}f})(g)\pi(1, \nu; g^{-1})\left\{P_\eta(\nu)\psi_{\eta,1}(k)+ \ll \text{other } K\text{-types} \gg\right\}dg \quad (35)$$

$$= \int_G (\overline{\pi_{sj}f})(g)\pi(1, \nu; g^{-1})\int_K \overline{\eta_{p1}}(k')\left\{P_\eta(\nu)\psi_{\eta,1}(kk')+ \ll \text{other } K\text{-types} \gg\right\}dk'dg$$

$$= P_\eta(\nu)\int_G (\overline{\pi_{sj}f})(g)\pi(1, \nu; g^{-1})\psi_{\eta,p}(k)dg$$

$$= P_\eta(\nu)(\overline{\pi_{sj}f})^\wedge(\psi_{\eta,p}; 1, \nu)(k)$$

Therefore, by substituting (35) for (33), it follows from Proposition 5.1 that

THEOREM 5.3 (REDUCTION FORMULA). Let us suppose that $\sigma \in \hat{M}$ and $\tau \in \hat{K}_\sigma$ satisfy (A). Let $f \in C^\infty_c(G)$. Then for each $i \in I_\tau$ and $\nu \in \mathcal{F}_c$

$$< S^\tau(\sigma, \nu + \sqrt{-1}\mu_\alpha)\hat{f}(\phi_{\tau,i,1}; \sigma, \nu + \sqrt{-1}\mu_\alpha)(g), v_{\tau,i} >$$

$$= \sum_{j,s \in I_\tau} \Pi_{ij}(g) \sum_{p \in I_\eta, \eta \in \hat{K}_{1,\tau}} C_{\tau,i,s;\eta,p} P_\eta(\nu)^{-1}$$

$$\times S^1(1, \nu)((-H_\alpha)^{n(\eta)} E^\eta_{p1}(\overline{\pi_{sj}f}))^\wedge(\psi_{1,1}; 1, \nu)(g).$$

We here note that

$$S^1(1, \nu)((-H_\alpha)^{n(\eta)} E^\eta_{p1}(\overline{\pi_{sj}f}))^\wedge(\psi_{1,1}; 1, \nu)(g) \quad (36)$$

is a W-invariant function of ν (see Lemma 2.4) and moreover, the product with $P_\eta(\nu)^{-1}$ is a holomorphic function of ν, because of the fact that $\overline{\pi_{sj}f} \in C^\infty_c(G)$ (see (35)).

5.2. We now apply the reduction formula in Theorem 5.3 to the Fourier inversion formula in Lemma 2.2. For each $\delta > 0$ and a sufficiently small $\epsilon > 0$ we put

$$\mathcal{F}_\epsilon(\delta) = \{\nu \in \mathcal{F}_c; 0 \le \text{Im}(\nu) \le \delta\} \cup \{\nu \in \mathcal{F}_c; |\nu(H_\alpha) - \sqrt{-1}\delta| \le \epsilon\}$$

and then, for $\mu \in \mathfrak{h}^*_c$ in Lemma 3.1 we define

$$\Gamma(\mu_\alpha) \text{ is the set of simple poles of } \mu(\sigma, \nu) \text{ in } \mathcal{F}_\epsilon(\mu(H_\alpha)) \text{ and} \quad (37)$$

$$R_{\sigma,\varsigma} \text{ is the residue of } \mu(\sigma, \nu) \text{ at } \nu = \varsigma \in \Gamma(\mu_\alpha).$$

Actually, it follows from the explicit form of the Plancherel measure $\mu(\sigma, \nu)$ (see [A1], §3 and [HC3], §36) that $\mu(\sigma, \nu)$ is a meromorphic function on $\mathcal{F}_c$ with nonzero simple poles on the imaginary axis and with polynomial growth as $\|\nu\|$ goes to ∞.

Let f be in $C^\infty_c(G)$. Then, as said in 2.4, $\hat{f}(\phi_{\tau,i,\alpha}; \sigma, \nu)(k)$ $((\nu, k) \in \mathcal{F}_c \times K)$ is a holomorphic function of uniform exponential type and, as a function of ν, $< S^\tau(\sigma(\alpha), \nu)$ $\hat{f}(\phi_{\tau,i,\alpha}; \sigma, \nu)(g), v_{\tau,i} >$ $(g \in G)$ is a holomorphic function of exponential type. Therefore, by changing the contour of the integration $\mathcal{F}$ in (16) for $\mathcal{F}_\epsilon + i\mu_\alpha$, the upper boundary of $F_\epsilon(\mu(H_\alpha))$, we see the following,

LEMMA 5.4.

$$f^P_{\tau,\sigma(\alpha)}(g) = \sum_{i \in I_\tau} \int_{\mathcal{F}_\epsilon} < S^\tau(\sigma(\alpha), \nu + \sqrt{-1}\mu_\alpha)\hat{f}(\phi_{\tau,i,\alpha}; \sigma, \nu + \sqrt{-1}\mu_\alpha)(g), v_{\tau,i} >$$

$$\times \, \mu(\sigma, \nu + \sqrt{-1}\mu_\alpha)d\nu$$

$$+ \sum_{i \in I_\tau} \sum_{\zeta \in \Gamma(\mu_\alpha)} R_{\sigma,\zeta} < S^\tau(\sigma, \zeta)\hat{f}(\phi_{\tau,i,\alpha}; \sigma, \zeta)(g), v_{\tau,i} > .$$

Especially, when $\alpha = 1(\sigma = \sigma(1))$, we can apply Theorem 5.3 to Lemma 5.4 and then we can deduce that

$$f^P_{\tau,\sigma}(g) = \sum_{i \in I_\tau} \sum_{j,s \in I_\tau} \Pi_{ij}(g) \sum_{p \in I_\eta, \eta \in \check{K}_{1,\tau}} C_{\tau,i,s;\eta,p}$$

$$\times \int_{\mathcal{F}_\epsilon} S^1(1, \nu)((-H_\alpha)^{n(\eta)} E^\eta_{p1}(\overline{\pi_{sj}}f))^\wedge(\psi_{1,1}; 1, \nu)(g)P_\eta(\nu)^{-1}\mu(\sigma, \nu + \sqrt{-1}\mu_\alpha)d\nu$$

$$+ \sum_{i \in I_\tau} \sum_{\zeta \in \Gamma(\mu_\alpha)} R_{\sigma,\zeta} < S^\tau(\sigma, \zeta)\hat{f}(\phi_{\tau,i,1}; \sigma, \zeta)(g), v_{\tau,i} > . \tag{38}$$

Although $P_\eta(\nu)^{-1}\mu(\sigma, \nu + i\mu_\alpha)$ is not W-invariant, $\mu(1, \nu)$ and

$$S^1(1, \nu)((-H_\alpha)^{n(\eta)} E^\eta_{p1}(\overline{\pi_{sj}}f))^\wedge(\psi_{1,1}; 1, \nu)(g)$$

are W-invariant (see [HC3], §36 and (36)). Therefore, since the product of the above function and $P_\eta(-\nu)^{-1} \mu(\sigma, -\nu + \sqrt{-1}\mu_\alpha)$ is holomorphic at $\nu = \sqrt{-1}\mu_\alpha$ (see the remark after Theorem 5.3), if we define a W-invariant function $Q(\eta; \sigma, \nu)$ as

$$Q(\eta; \sigma, \nu) = \frac{\{P_\eta(\nu)^{-1}\mu(\sigma, \nu + \sqrt{-1}\mu_\alpha) + P_\eta(-\nu)^{-1}\mu(\sigma, -\nu + \sqrt{-1}\mu_\alpha)\}}{2\mu(1, \nu)}, \tag{39}$$

we can deduce the following

THEOREM 5.5. Let us suppose that $\sigma \in \hat{M}$ and $\tau \in \hat{K}_\sigma$ satisfy (A). For $f \in C^\infty_c(G)$

$$f^P_{\tau,\sigma}(g) = \sum_{i \in I_\tau} \sum_{j,s \in I_\tau} \Pi_{ij}(g) \sum_{p \in I_\eta, \eta \in \check{K}_{1,\tau}} C_{\tau,i,s;\eta,p}$$

$$\times \int_{\mathcal{F}_\epsilon} S^1(1, \nu)((-H_\alpha)^{n(\eta)} E^\eta_{p1}(\overline{\pi_{sj}}f))^\wedge(\psi_{1,1}; 1, \nu)(g)Q(\eta; \sigma, \nu)\mu(1, \nu)d\nu$$

$$+ \sum_{i \in I_\tau} \sum_{\zeta \in \Gamma(\mu_\alpha)} R_{\sigma,\zeta} < S^\tau(\sigma, \zeta)\hat{f}(\phi_{\tau,i,1}; \sigma, \zeta)(g), v_{\tau,i} > .$$

We denote this formula as

$$f^P_{\tau,\sigma} = I_{\tau,\sigma}(f^P) + R_{\tau,\sigma}(f^P) \tag{40}$$

and call $I_{\tau,\sigma}(f^P)$ and $R_{\tau,\sigma}(f^P)$ the integral part and the residue part of $f^P_{\tau,\sigma}$ respectively.

References

[A1] Arthur, J.G., *Harmonic analysis of tempered distributions on semisimple Lie groups of real rank one*, Ph.D., Yale University (1970).

[A2] _____, *A Paley-Wiener theorem for reductive groups*, Acta Math. **150** (1983), 1-89.

[C] Campoli, O.A., *Paley-Wiener type theorems for rank-1 semisimple Lie groups*, Unión Matemática Argentina **29** (1980), 197-221.

[G] Gangolli, R., *On the Plancherel formula and the Paley-Wiener theorem for spherical functions on semisimple Lie groups*, Ann. of Math. **93** (1971), 150-165.

[H1] Helgason, S., *An analogue of the Paley-Wiener theorem for the Fourier transform on certain symmetric spaces*, Math. Ann. **165** (1966), 297-308.

[H2] _____, *A duality for symmetric spaces with applications to group representations*, Advances in Math. **5** (1970), 1-154.

[H3] _____, *A duality for symmetric spaces with applications to group representations, II. Differential equations and eigenspace representations*, Advances in Math. **22** (1976), 187-219.

[H4] _____, *Functions on symmetric spaces*, in "Harmonic Analysis on Homogeneous Spaces," Amer. Math. Soc. Providence, Rhode Island, 1973, pp. 101-146.

[HC1] _____, *Harmonic analysis on real reductive groups, I. The theory of the constant term*, J. Func. Anal. **19** (1975), 104-204.

[HC2] _____, *Harmonic analysis on real reductive groups, II. Wave packets in the Schwartz space*, Invent. Math. **36** (1976), 1-55.

[HC3] _____, *Harmonic analysis on real reductive groups, III. The Maass-Selberg relations and the Plancherel formula*, Ann. of Math. **104** (1976), 117-201.

[HC4] _____, *Two theorems on semi-simple Lie groups*, Ann. of Math. **83** (1966), 74-128.

[J] Johnson, K.D., *Composition series and intertwining operators for the spherical principal series. II*, Trans. Amer. Math. Soc. **215** (1976), 269-283.

[JW] Johnson, K.D. and N.R. Wallach, *Composition series and intertwining operators for the spherical principal series I*, Trans. Amer. Math. Soc. **229** (1977), 137-173.

[K1] Kawazoe, T., *An analogue of Paley-Wiener theorem on rank 1 semisimple Lie groups I*, Tokyo J. Math. **2** (1979), 397-407; *II*, Tokyo J. Math. **2** (1979), 409-421.

[K2] _____, *A transfrom on classical bounded symmetric domains associated with a holomorphic discrete series*, Tokyo J. Math. **12** (1989), 269-297.

[K3] _____, *Szegö operators and a Paley-Wiener theorem on $SU(1,1)$*, Keio Univ. Res. Rep. **5** (1990), 1-35.

[K4] _____, *A method of reduction in harmonic analysis on real rank 1 semisimple Lie groups II*, (preprint).

[Kn] Knapp, A.W., *Weyl group of a cuspidal parabolic*, Ann. Ec. Norm. Sup. (4)**8** (1975), 275-294.

[Ko] Kostant, B., *On the existence and irreducibility of certain series of representations*, Bull. Amer. Math. Soc. **75** (1969), 627-642.

[KS] Knapp, A.W. and E.M. Stein, *Intertwining operators for semisimple groups*, Ann. of Math. **93** (1971), 489-578.

[KW] Knapp, A.W. and N.R. Wallach, *Szegö kernels associated with discrete series*, Invent. Math **34** (1976), 163-200.

[V] Varadarajan, V.S., "Harmonic Analysis on Real Reductive Groups," Lecture Note in Math., Springer-Verlag, New York, 1977.

[Wa] Warner, G., "Harmonic Analysis on Semi-Simple Lie Groups II," Springer-Verlag, New York, 1972.

WAVELETS, SPLINE INTERPOLATION AND LIE GROUPS

Pierre Gilles LEMARIE

Université de Paris-Sud
Mathématiques
Bâtiment 425
91405 ORSAY Cedex
(France)

The purpose of this speech is to explain my construction of a so-called wavelet basis on stratified Lie groups [4]. I will first recall the classical notion of a wavelet basis, namely an orthonormal basis $\psi_{\epsilon,j,k}$ $(1 \leq \epsilon \leq 2^d - 1,\, j \in \mathbb{Z},\, k \in \mathbb{Z}^d)$ of $L^2(\mathbb{R}^d)$ generated from a finite number of (regular, oscillating and localized) functions ψ_ϵ by dyadic dilations and translations

$$\psi_{\epsilon,j,k}(x) = 2^{jd/2}\psi(2^j x - k).$$

I will show why we may expect such bases to be constructed on stratified Lie groups (where didlations and translations are defined with respect to the Lie structure), and especially how we can extend the notion of spline functions (which gave one of the first constructions of a wavelet basis on $\mathbb{R}$) to produce such a basis. I will restrict my speech to the simple case of the Heisenberg group to avoid notational burden.

I. Wavelet bases and the notion of multi-resolution analysis

A wavelet basis is an orthonormal basis of $L^2(\mathbb{R}^d)\psi_{\epsilon,j,k}$ $(1 \leq \epsilon \leq 2^d - 1,\, j \in \mathbb{Z},$ $k \in \mathbb{Z}^d)$ where :

(1) $$\psi_{\epsilon,j,k}(x) = 2^{jd/2}\psi_\epsilon(2^j x - k)$$

and the ψ_ϵ are a finite set of $2^d - 1$ functions which are :

- regular : for $\mid \alpha \mid \leq \alpha_0 \frac{\partial^\alpha}{\partial x^\alpha}\psi_\epsilon$ exists ;
- localized : the $\frac{\partial^\alpha}{\partial x^\alpha}\psi_\epsilon$ $(\mid \alpha \mid \leq \alpha_0)$ are rapidly decreasing at infinity ;
- oscillating : for $\mid \alpha \mid \leq \alpha_1,\, \int x^\alpha \psi_\epsilon(x)\,dx = 0$.

The coefficients $< f \mid \psi_{\epsilon,j,k} >$ then provide a local Fourier analysis of the function f, since the functions ψ_ϵ are localized both in space and frequency : the spatial localization is obvious and the frequential one comes from the flatness of the Fourier transform of ψ_ϵ both at 0 (oscillation) and at infinity (regularity). The coefficient $< f \mid \psi_{\epsilon,j,k} >$ then gives an information on the behaviour of f in the neighborhood of $\frac{k}{2^j}$ (roughly speaking, on a ball of radius $\frac{C}{2^j}$ around $\frac{k}{2^j}$) for the frequencies of order of magnitude 2^j (for frequencies ξ in an annulus $a2^j \leq \mid \xi \mid \leq b2^j$).

This notion of wavelet basis was introduced in 1985 by Y. Meyer [5], [7]. In 1986, S. Mallat introduced the notion of multi-resolution analysis as a powerful tool both to describe and to construct wavelet bases [6].

A multi-resolution analysis is a sequence V_j, $j \in \mathbb{Z}$, of closed subspaces of $L^2(\mathbb{R}^d)$ such that :

(2.1) $V_j \subset V_{j+1}$, $\bigcap_{j \in \mathbb{Z}} V_j = \{0\}$, $\bigcup_{j \in \mathbb{Z}} V_j$ is dense in $L^2(\mathbb{R}^d)$;

(2.2) $f(x) \in V_j \Leftrightarrow f(2x) \in V_{j+1}$ (dilation invariance) ;

(2.2) V_0 has a Riesz basis $(g(x-k))_{k \in \mathbb{Z}^d}$ with g localized and regular (translation invariance).

The first property expresses that the sequence V_j is an approximation sequence for L^2 ; more precisely, if W_j is the orthogonal complement of V_j in V_{j+1} (i.e. $V_{j+1} = V_j \oplus W_j$), we have the following decomposition of $L^2(\mathbb{R}^d)$:

$$(3) \qquad L^2 = \bigoplus_{j \in \mathbb{Z}} W_j.$$

The point is that, given a multi-resolution analysis, we can construct a wavelet basis associated to it. More precisely, we have :

THEOREM.

a) There exists a function φ in V_0, localized and regular, such that the $\varphi(x-k)$, $k \in \mathbb{Z}^d$, are an orthonormal basis of V_0.

b) There exist $2^d - 1$ functions ψ_ϵ in W_0, localized, regular and oscillating, such that the $\psi_\epsilon(x-k)$, $1 \leq \epsilon \leq 2^d - 1$, $k \in \mathbb{Z}^d$, are an orthonormal basis of W_0.

c) As a corollary, the $\psi_{\epsilon,j,k}$ are a wavelet basis in $L^2(\mathbb{R}^d)$.

The proof and the consequences of this theorem are fully developped in the book of Yves Meyer [8]. Since V_0 and W_0 are subspaces of V_1, we have that the Fourier transforms $\hat{\varphi}$ of φ and $\hat{\psi}_\epsilon$ of ψ_ϵ satisfy, for $2\pi\mathbb{Z}^d$-periodic functions m_0 and m_ϵ :

$$(4.1) \qquad \hat{\varphi}(\xi) = m_0\left(\frac{\xi}{2}\right)\hat{\varphi}\left(\frac{\xi}{2}\right)$$

$$(4.2) \qquad \hat{\psi}_\epsilon(\xi) = m_\epsilon\left(\frac{\xi}{2}\right)\hat{\varphi}\left(\frac{\xi}{2}\right)$$

$$(4.3) \qquad \text{the matrix } \left(m_\epsilon\left(\frac{\xi}{2} + k\pi\right)\right), \ 0 \leq \epsilon \leq 2^d - 1, \ k \in \{0,1\}^d, \text{ is unitary.}$$

As we shall see below, the function φ is easily determined, and hence the function $m_0(\xi) = \sum_{k \in \mathbb{Z}^d} < \frac{1}{2}\varphi(\frac{x}{2}) \mid \varphi(x-k) > e^{-ik \cdot \xi}$. The problem is to construct the $2^d - 1$ remaining functions m_ϵ and was solved in 1987 by K. Grochenig. However, there are three easy cases where the ψ_ϵ can be directly constructed :

* the $1 - D$ case : in that case, we have the solution $m_1(\xi) = e^{-i\xi}\overline{m_0}(\xi + \pi)$;
* the tensor-product case : in case that $V_0^{(d)} \subset L^2(\mathbb{R}^d)$ is the closure in $L^2(\mathbb{R}^d)$ of a tensor product $V_0^{(1)} \otimes \cdots \otimes V_0^{(1)}$ where $V_0^{(1)}$ is associated with a multi-resolution of $L^2(\mathbb{R})$ with basic functions φ and ψ, then the function $\varphi \otimes \cdots \otimes \varphi$ provides through $\mathbb{Z}^d$-translations an orthonormal basis of $V_0^{(d)}$ whereas the functions ψ_ϵ defined as

(5) $\qquad \psi_\epsilon(x) = \psi^{(\epsilon_1)}(x_1) \cdots \psi^{(\epsilon_d)}(x_d), \quad \psi^{(0)} = \varphi, \quad \psi^{(1)} = \psi, \quad \epsilon \neq (0,\cdots,0)$

are wavelets associated to $W_0^{(d)}$.

* the third easy case, where the Riesz basis $g(x - k)$ of V_0 is given by an interpolant function g for $\mathbb{Z}^d$, is developped in the next section.

II. Self-similar interpolants and wavelets

Let g be a self-similar interpolant function for $\mathbb{Z}^d$:

(6.1) $\qquad\qquad g(0) = 1, \quad g(k) = 0 \text{ for } k \in \mathbb{Z}^d\backslash\{0\} \quad \text{(interpolation)}$

(6.2) $\qquad\qquad g\left(\frac{x}{2}\right) = g(x) + \sum_{k \in \mathbb{Z}^d, \frac{k}{2} \notin \mathbb{Z}^d} g\left(\frac{k}{2}\right)g(x - k)$

and such that g is regular and localized :

(6.3) $\qquad\qquad \text{for } |\alpha| \leq \alpha_0, \quad \left|\frac{\partial^\alpha g}{\partial x^\alpha}(x)\right| \leq C \exp - D \|x\|.$

It can then be proved that :

(6.4) $\qquad\qquad \text{for } |\alpha| \leq \alpha_0, \quad \sum_{k \in \mathbb{Z}^d} k^\alpha g(x - k) = x^\alpha.$

Moreover the $g(x - k)$, $k \in \mathbb{Z}^d$, are a Riesz basis for a space V_0, which generates by dyadic dilations a multi-resolution analysis for $L^2(\mathbb{R}^d)$.

We can compute φ directly by observing that, since the $\varphi(x - k)$ are an orthonormal set, the Fourier transform $\hat{\varphi}$ of φ satisfies

(7) $\qquad\qquad\qquad \sum_{k \in \mathbb{Z}^d} |\hat{\varphi}(\xi + 2k\pi)|^2 = 1$

so that we have a solution φ given as :

$$(8) \qquad \hat{\varphi}(\xi) = \frac{\hat{g}(\xi)}{[\sum_{k \in \mathbb{Z}^d} | \hat{g}(\xi + 2k\pi) |^2]^{1/2}}.$$

However, we will prefer to compute φ by an other way, namely by performing a so-called "Gram orthonormalization". If H is a Hilbert space and $(e_\alpha)_{\alpha \in \mathcal{A}}$ a Riesz basis of H, we can form the Gram matrix $G = (< e_\alpha / e_\beta >)_{\alpha, \beta \in \mathcal{A}}$ and then compute $G^{-1/2} = (\mu_{\alpha, \beta})_{\alpha, \beta \in \mathcal{A}}$; we then have an orthonormal basis of H with $\varphi_\alpha = \sum_{\beta \in \mathcal{A}} \mu_{\alpha, \beta} e_\beta$, $\alpha \in \mathcal{A}$. In case of the family $g(x - k)$, we obtain of course a familly φ_k with $\varphi_k(x) = \varphi_0(x - k)$, but this construction can (and will) be applied to more general cases where the Fourier transform doesn't fit.

By instance, we can now easily construct the wavelets ψ_ϵ. We have a supplement X_0 of V_0 in V_1 by considering the closed linear span of the $g(2x - k)$, $k \in \mathbb{Z}^d$ and $\frac{k}{2} \notin \mathbb{Z}^d$, since for $f \in V_1$ we can write $f(x) = \sum_{k \in \mathbb{Z}^d} f(k)g(x - k) + r(x)$ and

$$r(x) = \sum_{k \in \mathbb{Z}^d, \frac{k}{2} \notin \mathbb{Z}^d} r(\frac{k}{2})g(2x - k).$$

If P_0 is the orthogonal projection on V_0, we then have a Riesz basis of W_0 just by taking up the $(Id - P_0)[g(2x - k)]$ $(k \in \mathbb{Z}^d, \frac{k}{2} \notin \mathbb{Z}^d)$; a Gram orthonormalization provides then an orthonormal basis ψ_k $(k \in \mathbb{Z}^d, \frac{k}{2} \notin \mathbb{Z}^d)$. It is very easy to verify that for $k' \in \mathbb{Z}^d$, $\psi_k(x - k') = \psi_{k+2k'}(x)$, so that this basis is generated through $\mathbb{Z}^d$-translations from a finite set of $2^d - 1$ functions ψ_ϵ. We will see that the ψ_ϵ are (exponentially) localized, so that we have regularity from (6.3) and oscillation from (6.4).

For the localization, we have the following lemma :

LEMMA. Let t be a subset of $\mathbb{Z}^d$ and $M = (m_{t,s})_{t,s \in T}$ a matrix on T such that :

$$| m_{t,s} | \leq C \exp - D | s - t |$$

and such that M is invertible as an operator on $\ell^2(T)$. Then $M^{-1} = (\mu_{t,s})_{t,s \in T}$ satisfies :

$$| \mu_{t,s} | \leq C' \exp - D' | s - t |.$$

COROLLARY. If $X \subset L^2$ is the closed linear span of a familly $g_t(x)$, $t \in T \subset \mathbb{Z}^d$, such that $| g_t(x) | \leq C \exp - D | x - t |$ and such that the g_t are a Riesz basis of X, then the Gram orthonormalization provides an orthonormal basis ψ_t, $t \in T$, with

$$| \psi_t(x) | \leq C' \exp - D' | x - t |.$$

The corollary can be easily deduced from the lemma, since we have

$$G^{-1/2} = \frac{2}{\pi} \int_0^\infty (Id + t^2 G)^{-1} dt.$$

III. A fundamental example : spline functions

Let X be a quasi-uniform subdivision of $\mathbb{R}$, id est $X = (\cdots < x_k < x_{k+1} < \cdots)_{k \in \mathbb{Z}}$ with $\inf_{k \in \mathbb{Z}} |x_k - x_{k+1}| > 0$ and $\sup_{k \in \mathbb{Z}} |x_k - x_{k+1}| < +\infty$. Then a square-integrable spline function of order N (or degree $N-1$) with modes in X is a function $f \in L^2$ which satisfies one of the following equivalent assertions :

(9.1) f is of class C^{N-2} and its restriction to any interval (x_k, x_{k+1}) is polynomial with degree $\leq N - 1$.

(9.2) the N-th derivative of f is an (infinite) sum of Dirac masses located in X : $f^{(N)} = \sum_{x \in X} \alpha_x \delta_x$.

(9.3) f is orthogonal to any $g^{(N)}$, where $g \in H^N$ satisfies $g(x) = 0$ for all x in X

(9.4) (in case N is even, $N = 2p$) f is in H^p and has the following minimization property : if $F \in H^p$ satisfies $F(x) = f(x)$ for all x in X then $\| F^{(p)} \|_2 \geq \| f^{(p)} \|_2$.

The minimizing problem in (9.4) is well-posed since we have a Poincaré inequality :

$$\| F \|_{H^p} \sim \| F^{(p)} \|_2 + \left(\sum_{x \in X} | F(x) |^2 \right)^{1/2}.$$

The same inequality shows that, in (9.3), the space of the $g^{(N)}$, with $g \in H^N$ and $g(x) = 0$ for x in X, is closed in L^2 and hence is exactly the orthogonal of the spline functions.

We can now have a multi-resolution analysis for $L^2(\mathbb{R})$ by taking V_j as the set of square-integrable spline functions of order N with modes in $\frac{1}{2^j}\mathbb{Z}$. We have a Riesz basis $g(x - k)$ of V_0, where $g = \chi_{[0,1]} * \cdots * \chi_{[0,1]} = \chi_{[0,1]}^{(*)N}$ and $\chi_{[0,1]}$ is the characteristic function of $[0, 1]$. We then obtain a spline wavelet ψ with exponential decay [3]. Of course, ψ is of class C^{N-2}. Moreover it is oscillating since $\psi \in V_0^\perp$ and we know that a function in $V_0^\perp$ is of the form $g^{(N)}$ with g in H^N.

There are two other noticeable constructions of a wavelet spline :

* (G. Battle [1]) : define L as a spline of order $2N$ with modes in $\mathbb{Z}$ such that $L(0) = 1$, $L(k) = 0$ for $k \neq 0$, and define Λ as $\Lambda = L^{(N)}(2x - 1)$. Then the $\Lambda(x - k)$, $k \in \mathbb{Z}$, are a Riesz basis in W_0.

* (J. O. Strömberg [9]) : define L as a spline of order $2N$ with modes in $\mathbb{Z} \cup \{\frac{1}{2} + \mathbb{N}\}$ such that $L(\frac{1}{2}) = 1$ and L is 0 at the other modes. Put $\psi = \frac{L^{(N)}}{\|L^{(N)}\|_2}$; then the $\psi(x - k)$ are an orthonormal basis in W_0.

Classically, we can extend the notion of spline to $\mathbb{R}^d$ by making use of the tensor product. We then have a multi-resolution analysis and wavelet bases for $L^2(\mathbb{R}^d)$. But I'd like to point out another possible extension of the notion of spline.

Let X be a set in $\mathbb{R}^d$ such that $\inf_{x,x' \in X, x \neq x'} \| x - x' \| > 0$ and $\sup_{y \in \mathbb{R}^d} d(y, X) < +\infty$. A spline-surface with modes in $X \subset \mathbb{R}^d$ and of degree $s > d$ will be a function f in $H^{s/2}$ such that $(-\Delta)^{s/2}f$ is a sum of Dirac masses at X : $(-\Delta)^{s/2}f = \sum_{x \in X} \alpha_x \delta_x$.

If $X = \frac{1}{2^j}\mathbb{Z}^d$, we obtain a space V_j of a multi-resolution analysis for $L^2(\mathbb{R}^d)$. The point is that the operator $(-\Delta)^{s/2}$ is well-adapted for defining a multi-resolution analysis since it is both invariant for translations and homogeneous for dilations. Moreover, $-\Delta$ is elliptic and therefore our spline surfaces are regular. At last, $V_0^\perp$ will be the set of functions of the form $(-\Delta)^{s/2}g$, with $g \in H^s$ and $g(k) = 0$ for $k \in \mathbb{Z}^d$; hence the functions in $V_0^\perp$ are oscillating.

If s is an even integer, then we can easily show, through Fourier series, that there exists a spline-surface L such that $L(0) = 1$ and $L(k) = 0$ for $k \neq 0$ and that this spline has exponential decay. Hence, we obtain (regular, localized and oscillating) wavelets for $L^2(\mathbb{R}^d)$.

IV. Analysis on Heisenberg group

The Heisenberg group can be defined as a real nilpotent simply connected Lie group, with a basis X, Y, Z of its Lie algebra such that $[X,Y] = 2Z$. Many of the notions we have used in our wavelet theory can be used in the context of the Heisenberg group. For instance, we have the following "dictionary" between $\mathbb{R}^d$ and $\mathbb{H}$:

	$\mathbb{R}^d$	$\mathbb{H}$
element:	$X = (x_1, \cdots, x_d)$	$h = (x, y, z)$
translation:	$X + X' = (x_1 + x_1', \cdots, x_d + x_d')$	$h'h = (x + x', y + y', z + z' + x'y - y'x)$

		(left-translation of h by h')
dilation:	$tX = (tx_1, \cdots, tx_d)$	$t \cdot h = (tx, ty, t^2 z)$
		(distributivity $t \cdot (h'h) = (t.h')(t.h)$)
Haar measure:	$dX = dx_1 \cdots dx_d$	$dh = dx\,dy\,dz$
	$d(tX) = t^d dX$	$d(t \cdot h) = t^4 dh$
distance and norm	$\| X - Y \|$	$\| h^{-1} k \|$
	$\| X \| = \left(\sum_i |x_i|^2 \right)^{1/2}$	$\| h \| = (x^4 + y^4 + z^2)^{1/4}$
	$\| tX \| = t \| X \| \quad (t > 0)$	$\| t \cdot h \| = t \| h \| \quad (t > 0)$
Laplacian	Laplacian (elliptic operator)	Sublaplacian (hypoelliptic operator)
	$-\Delta = -\sum_{i=1}^d \frac{\partial^2}{\partial x_i^2}$	$J = -X^2 - Y^2, \ X = \frac{\partial}{\partial x} - y\frac{\partial}{\partial z},$
		$Y = \frac{\partial}{\partial y} + x\frac{\partial}{\partial z}$
	$(-\Delta)(f(X - X_0)) =$	$J(f(h_0^{-1}h)) = (Jf)(h_0^{-1}h)$
	$\quad = (-\Delta f)(X - X_0)$	
	$(-\Delta(f(tX)) = t^2(-\Delta f)(tX)$	$J(f(t \cdot h)) = t^2(Jf)(t \cdot h)$
Sobolev spaces	$H^s = \mathcal{D}((-\Delta)^{s/2}) \ \ s \geq 2$	$H^s = \mathcal{D}(J^{s/2}) \ \ s \geq 0$
Discrete subgroup	$\mathbb{Z}^d$	$\mathbb{Z}^3 = \mathbb{Z}X \oplus \mathbb{Z}Y \oplus \mathbb{Z}Z$
	$2\mathbb{Z}^d \subset \mathbb{Z}^d$	$2 \cdot \mathbb{Z}^3 \subset \mathbb{Z}^3$

At that point, we may ask whether there are wavelet bases on $\mathbb{H}$, i.e. orthonormal bases $\psi_{\epsilon,j,k}$ ($1 \leq \epsilon \leq 2^4 - 1, \ j \in \mathbb{Z}, \ k \in \mathbb{Z}^3$) generated from a finite set of 15 regular, localized and oscillating functions ψ_ϵ through "Heisenberg dilations and translations" :

(10) $\quad \psi_{\epsilon,j,k} = 2^{2j}\psi_\epsilon(k^{-1}(2^j.h)) = 4^j\psi_\epsilon(2^j x - k_1, 2^j y - k_2, 2^{2j} z - k_3 + 2^j k_2 x - 2^j k_1 y).$

THEOREM [4]. *For every integer N, there exist 15 functions ψ_ϵ on $\mathbb{H}$ such that the $\psi_{\epsilon,j,k}$ are an orthonormal basis for $L^2(\mathbb{H})$ and the ψ_ϵ satisfy :*

(i) *for every left-invariant differential operator $P(x, D)$ homogeneous of degree α (with respect to Heisenberg dilations) with $\alpha \leq N$, $P(x, D)\psi_\epsilon$ exists (regularity) and is exponentially decreasing at infinity (localization) :*

$$| P(x, D)\psi_\epsilon(h) | \leq C \ exp - D \| h \|_{\mathbb{H}}$$

(ii) *for every polynomial P homogenous of degree α (with respect to Heisenberg dilations) with $\alpha \leq N$, $\int P(h)\psi_\epsilon(h)dh = 0$ (oscillation).*

V. Construction of wavelet bases :
generalized spline surfaces on Heisenberg group.

In order to produce wavelet bases, we will mimic on the Heisenberg group the general construction of wavelets on $\mathbb{R}^d$, carefully avoiding any reference to the Fourier transform (which might be surprising for a speech in a conference on harmonic analysis...) and using only "geometrical" tools.

First of all, we will produce a self-similar interpolant for $\mathbb{Z}^3$ (and hence a multi-resolution analysis for $L^2(\mathbb{H})$). By construction, this interpolant will be regular and the forthcoming wavelets will be regular and oscillating. What will be very uneasy to obtain, however, will be the localization of the interpolant. But, as a matter of fact, we will obtain exponential decay for this interpolant and its first derivatives, and this exponential localization will not be losen in the Gram orthonormalization processes.

Now, let us introduce our self-similar interpolant L. For X a "nice set" in $\mathbb{H}$ (more precisely, we ask that $\inf_{x,x' \in X, x \neq x'} \| x^{-1}x' \| > 0$ and $\sup_{h \in \mathbb{H}} \inf_{x \in X} \| x^{-1}h \| < +\infty$) and $N > 4$, we define a generalized spline surface with modes in X and of degree N as a function $g \in H^{N/2}$ such that $J^{N/2}g$ is a sum of Dirac masses in X. We can easily prove that spline surfaces exists and the map $g \rightarrow (g(x))_{x \in X}$ is a bijection between the space of spline surfaces (of degree N) with modes in X and $\ell^2(X)$; it follows from a Poincaré inequality on $H^{N/2}$:

$$ \| g \|_{H^{N/2}} \sim \| J^{N/4}g \|_2 + \left(\sum_{x \in X} | g(x) |^2 \right)^{1/2} . $$

(A slight mistake has been made in the proof of this inequality in [4] but the proof can be easily modified in order to be correct). As a corollary, we obtain that the orthogonal in $L^2(\mathbb{H})$ of the space of spline-surface is the set of the functions of the form $J^{N/2}G$ with $G \in H^N$ and $G(x) = 0$ for all $x \in X$. We may now define L as the (unique) spline-surface of degree N with nodes in $\mathbb{Z}^3$ such that $L(0) = 1$ and $L(k) = 0$ for all $k \in \mathbb{Z}^3 \backslash \{0\}$. The space V_0 spanned by the $L(k^{-1}h)$, $k \in \mathbb{Z}^3$, is just the space of spline surfaces with nodes in $\mathbb{Z}^3$. Hence the wavelets ψ_ϵ we will produce by Gram orthonormalization will be regular (since $\psi_\epsilon \in V_{-1} \subset H^{N/2}$; in particular, $P(x,D)\psi_\epsilon$ is continuous if $P(x,D)$ is left-invariant and homogeneous of degree α with $\alpha < \frac{N}{2} - 2$) and oscillating (since $\psi_\epsilon \in V_0^\perp$ so that there exists $G_\epsilon \in H^{N/2}$ such that $\psi_\epsilon = J^{N/2}G_\epsilon$; moreover we will be able to prove that G_ϵ is exponentially decreasing so that $\int P(h)\psi_\epsilon(h)dh = \int G_\epsilon(h)J^{N/2}P(h)dh = 0$ if P is homogeneous of degree $\alpha < N$).

So we can see that the only point to prove is the localization of L. This is done in several steps for N an even integer :

Step 1. Approximation by compactly supported functions. In order to prove that L has exponential decay, we will decompose L into $L = \sum_1^\infty \Lambda_j$ with Supp $\Lambda_j \subset \mathcal{B}(0, C_1 j)$ and $\| \Lambda_j \|_\infty \leq C_2 \epsilon^j$ ($0 < \epsilon < 1$). L is the solution of the minimizing problem : $L(0) = 1$,

$L(k) = 0$ for $h \in \mathbb{Z}^3 \backslash \{0\}$ and $\| J^{N/4} L \|_2$ minimum. Let us define L_j as the solution of the minimizing problem : $L_j(0) = 1$, $L_j(k) = 0$ for $k \in \mathbb{Z}^3 \backslash \{0\}$, $L_j = 0$ outside the ball $B(0, C_1 j)$ and $\| J^{N/4} L_j \|_2$ minimum. Then L_j is a compactly supported approximation of L in the sense that $L_j \to L$ in $H^{N/4}$ as $j \to +\infty$, so that we can decompose L as $L = L_1 + \sum_{j=2}^{\infty} (L_j - L_{j-1})$. But, since L_j and L_{j-1} have the same values on $\mathbb{Z}^3$, we can replace the norm $\| L_j - L_{j-1} \|_{H^{N/4}}$ by $\| J^{N/4} (L_j - L_{j-1}) \|_2$ (from the Poincaré inequality) and, from the minimizing property of L_j, we have that $< J^{N/4} (L_j - L_{j-1}) \mid J^{N/4} L_{j+p} >= 0$ if $p \geq 0$ so that : $\| \sum_{j=2}^{\infty} (L_j - L_{j-1}) \|_{H^{N/2}} \sim \left(\sum_{j=2}^{\infty} \| J^{N/4} (L_j - L_{j-1}) \|_2^2 \right)^{1/2}$. But, again from the minimizing property of L_j and since $J^{N/2}$ is a differential, hence local operator for N even, we have for $\psi_j \in C_0^{\infty}(\mathbb{H})$, $\psi_j \equiv 0$ outside $B(0, C_1 j)$ and $\equiv 1$ in the neighborhood of $B(0, C_1(j-1))$:

$$< J^{N/4}(f_{j+1} - f_j) \mid J^{N/4}(f_{j+1} - f_j) >=< [J^{N/2}, (1 - \psi_j)](f_{j+1} - f_j) \mid f_{j-1} - f_j >$$

so that, by a careful choice of ψ_j, we obtain :

$$\| f_{j+1} - f_j \|_{H^{N/2}} \leq \epsilon \| f_j - f_{j-1} \|_{H^{N/2}} .$$

This proof gives an exponential control over the derivatives of L up to any order $< N/2 - 2$; however we can prove much better and obtain that L is in $H^{N-2-\epsilon}$ and that its derivatives are controlled exponentially up to any order $< N - 4$.

Step 2. Control of $J^{N/2} L$. We know that $J^{N/2} L$ is of the form $J^{N/2} L = \sum a_k \delta_k$. Moreover, if $\varphi \in C_0^{\infty}$ is such that $\varphi(0) = 1$, $\varphi(k) = 0$ for $k \in \mathbb{Z}^3$, then $a_k =< J^{N/2} L \mid \varphi(k^{-1} h) >=< L \mid J^{N/2} \varphi(k^{-1} h) >$, hence the sequence a_k has exponential decay.

Step 3. Interpolation between L and $J^{N/2} L$. From the exponential control of both L and $J^{N/2} L$, we can obtain the exponential control of the derivatives of L of order $< N - 4$ by decomposing L on... a very regular wavelet basis ! (This is a good example of an application of wavelet bases).

VI. Extensions of the construction

Of course, this construction can be made on other Lie groups that the Heisenberg group. But not on any Lie group, however. First of all, if we want to have dilations on the group, the group has to be a nilpotent one. Secondly, if we want to use a sublaplacian, the group has to be stratified : the Lie algebra $\underline{n}$ of the group can be decomposed as

$$\underline{n} = \overset{r}{\underset{i=1}{\oplus}} N_i \text{ with } [N_1, N_i] = N_{i+1} \text{ (and } N_{r+1} = \{0\}).$$

(Maybe, if the group were not stratified, we could use the Rockland operator instead of the sublaplacian). Thirdly, if we want to have a discrete subgroup stable by a dilation by 2 and such that the union of its dyadic dilated versions is dense in the whole group, we have an algebraic condition to be fulfilled : the constant of structure of the group should be rational.

If N is a real nilpotent simply connected stradified Lie group of homogeneous dimension Q (where $d(t \cdot x) = t^Q dx$, i.e. $Q = \sum_{i=1}^{r} i \dim N_i$) and if N has rational constants of structure, then $L^2(N)$ has wavelet bases, i.e. orthonormal bases generated through dyadic translations and dilations from a finite set of $2^Q - 1$ regular, localized and oscillating functions.

If the constants are irrational, we have no good subgroup and we cannot expect to produce a basis generated from a finite set of functions. But, we may consider in each $N_i = \mathbb{R}^{\dim N_i}$ the subset $\mathbb{Z}_i = \mathbb{Z}^{\dim N_i}$ and $Z = \oplus \mathbb{Z}_i$ and define a "generalized multi-resolution analysis" V_j through the splines of order N with modes in $\frac{1}{2^j}Z$. Of course, V_0 has no Riesz basis of the form $g(k^{-1}h)$, $k \in Z$, since Z is not a subgroup. But it has a basis g_k such that $g_k(k') = \delta_{k,k'}$ and with exponential control : for $P(x, D)$ left-invariant and of degree $< 2N - Q$, $\| P(x, D)g_k(h) \| \leq C \exp - D \mid k^{-1}h \mid$. We then obtain a basis $\psi_{\epsilon,j,k}$ of L^2 composed with regular, oscillating and localized functionc $\psi_{\epsilon,j,k}$ such that the functions $2^{-jQ/2}\psi_{\epsilon,j,k}(2^{-j} \cdot (kh))$ are uniformly controlled : $\mid P(x, D)\{2^{-jQ/2}\psi_{\epsilon,j,k}(2^{-j} \cdot (kh))\} \mid \leq C \exp - D \| h \|$.

That uniform control is of course of great use in analysis. We can mimic such a construction on a compact Riemann manifold [6], through we have neither translations nor dilations, by using splines defined with help of the Laplace-Beltrami operator and with nodes in a "nice set". Again, we produce bases with "uniform" exponential control.

References

[1] BATTLE, G. *A block spin construction of ondelettes.* Comm. Math. Phys. 110 (1987), 601-615.

[2] JAFFARD, S. Thèse, Ecole Polytechnique, 1989.

[3] LEMARIE, P. G. *Ondelettes à localisation exponentielle.* J. Math. Pures et Appl. 67 (1988), 227-236.

[4] LEMARIE, P. G. *Bases d'ondelettes sur les groupes de Lie stratifiés.* Bull. Soc. Math. France 117 (1989), 211-232.

[5] LEMARIE, P. G. & MEYER, Y. *Ondelettes et bases hilbertiennes.* Rev. Mat. Iberoamericana 2 (1986), 1-17.

[6] MALLAT, S. *A theory for multi-resolution signal decomposition : the wavelet representation.* IEEE Trans. PAMI 11 (1989), 674-693.

[7] MEYER, Y. *Principe d'incertitude, bases hilbertiennes et algèbres d'opérateurs.* Sém. Bourbaki, fév. 1986.

[8] MEYER, Y. *Ondelettes et opérateurs.* Tome 1. Paris, Hermann, 1990.

[9] STROMBERG, J. O. *A modified Franklin system and higher-order systems in* $\mathbb{R}^n$, in Conf. on Harmonic Analysis in honor of A. Zygmund, vol. 2, 1988, 475-494.

PRINCIPAL VALUES OF CAUCHY INTEGRALS, RECTIFIABLE MEASURES AND SETS

Pertti Mattila

Department of Mathematics, University of Jyväskylä

P.O. Box 35, SF–40351 Jyväskylä, Finland

The extensive studies started by A. P. Calderón in the sixties and continued by many authors up to-day have revealed that the Cauchy integrals

$$C_\Gamma f(z) = \int_\Gamma \frac{f(\zeta)\, d\zeta}{\zeta - z}$$

behave very well on sufficiently regular, not necessarily smooth, curves Γ, see [CCFJR], [D] and [MT]. Thus for example if Γ is a rectifiable arc and $f \in L^1(\Gamma)$, $C_\Gamma f(z)$ exists almost everywhere on Γ as a principal value. Below we shall discuss converse questions by replacing $f\, ds$ on Γ by general finite Borel measures μ on $\mathbb{C}$. Thus we ask whether the existence

$$\exists\, C_\mu(z) = \lim_{\varepsilon \downarrow 0} \int_{|\zeta - z| \geq \varepsilon} (\zeta - z)^{-1}\, d\mu\, \zeta \in \mathbb{C} \tag{1}$$

of a finite principal value for μ almost all $z \in \mathbb{C}$ would force μ to be rectifiable in some sense. Obviously there are many measures far from rectifiable for which the integral converges absolutely, and so the principal value exists. For example, μ could be the Lebesgue measure on some set of positive and finite area. μ could be also a natural measure on some fractal curve, like the von Koch snowflake, satisfying $\mu B(z,r) \leq c\, r^s$ for some $s > 1$, whence $\int |\zeta - z|^{-1}\, d\mu\zeta < \infty$. Here $B(z,r)$ is the closed disc with center z and radius r. To exclude such "more than one-dimensional" measures, we shall impose the assumption

$$\liminf_{r \downarrow 0} \mu B(z,r)/r > 0 \tag{2}$$

for μ almost all $z \in \mathbb{C}$. The following results were established in [MP2]:

Theorem 1. *Let μ be a finite non-negative Borel measure on $\mathbb{C}$ satisfying (1) and (2) for μ almost all $z \in \mathbb{C}$. Then μ is rectifiable in the sense that there are rectifiable curves $\Gamma_1, \Gamma_2, \ldots$ such that*

$$\mu\left(\mathbb{C} \setminus \bigcup_{i=1}^{\infty} \Gamma_i\right) = 0.$$

The same conclusion holds for complex Borel measures replacing the assumption (2) by the condition that μ is absolutely continuous with bounded Radon–Nikodym derivative with respect to the one-dimensional Hausdorff measure $\mathcal{H}^1$ on an $\mathcal{H}^1$ measurable set E such that $\mathcal{H}^1(E) < \infty$ and

$$\liminf_{r \downarrow 0} \mathcal{H}^1\big(E \cap B(z,r)\big)/r > 0 \tag{3}$$

for $\mathcal{H}^1$ almost all $z \in E$. This is essentially the content of the following theorem.

Theorem 2. *Let E be as above and let $\varphi\colon E \to \mathbb{C} \setminus \{0\}$ be a bounded $\mathcal{H}^1$ measurable function. If the limit*

$$\lim_{\varepsilon \downarrow 0} \int_{E \setminus B(z,\varepsilon)} (\zeta - z)^{-1} \varphi(\zeta) \, d\mathcal{H}^1 \zeta$$

exists and is finite for $\mathcal{H}^1$ almost all $z \in E$, then E is rectifiable in the sense that there are rectifiable curves $\Gamma_1, \Gamma_2, \ldots$ such that

$$\mathcal{H}^1 \left(E \setminus \bigcup_{i=1}^{\infty} \Gamma_i \right) = 0.$$

This definition of rectifiability of E is equivalent to Besicovitch's regularity discussed in [**FK**]. Note that there are plenty of sets E with $0 < \mathcal{H}^1(E) < \infty$ satisfying (3) which are purely unrectifiable in the sense that they meet every rectifiable curve in a set of length zero. For example, such are the self-similar 1-sets, which are not line segments, considered in [**FK**] and [**MP1**].

When combined with the aforementioned results on the existence of principal values on rectifiable arcs (C^1 is enough here), Theorem 2 leads to the following characterization of the rectifiability of E:

Theorem 3. *Let E be an $\mathcal{H}^1$ measurable subset of $\mathbb{C}$ with $\mathcal{H}^1(E) < \infty$. Then E is rectifiable (in the sense of Theorem 2) if and only if (3) holds for $\mathcal{H}^1$ almost all $z \in E$ and E can be written as $E = \cup_{i=1}^{\infty} E_i$ such that for all i the limit*

$$\lim_{\varepsilon \downarrow 0} \int_{E_i \setminus B(z,\varepsilon)} (\zeta - z)^{-1} \, d\mathcal{H}^1 \zeta$$

exists and is finite for $\mathcal{H}^1$ almost all $z \in E_i$.

I don't know whether the countable decomposition of E here is necessary or not. That is, I don't know whether the principal value exists $\mathcal{H}^1$ almost everywhere on E for the rectifiable set E itself.

One motivation for studying questions to this direction comes from complex analysis. Often one can represent a given complex analytic function f outside a compact set E as a Cauchy transform with respect to a measure μ supported on E;

$$f(z) = \int \frac{d\mu\zeta}{\zeta - z}.$$

Geometric properties of E are then reflected in μ and from this fact one would like to relate them with the behavior of f. For example, suppose that $\mathcal{H}^1(E) < \infty$ and $\mathcal{H}^1(E \cap \Gamma) = 0$ for every rectifiable curve Γ. It is not known whether such an E has zero analytic capacity. The positive answer would be equivalent to showing that E supports no finite non-zero Borel measure μ such that the Cauchy transform $z \mapsto \int (\zeta - z)^{-1} \, d\mu\zeta$ is bounded in $\mathbb{C} \setminus E$. See e.g. [**FX**], [**G**], [**J**], [**K**], [**MP1**] and [**MT**] for such and other questions on complex and harmonic analysis related to Cauchy transforms of measures.

We shall now describe the basic ideas for the proofs of Theorems 1 and 2, and we shall discuss some related results obtained in connection of them. For more details, see [**MP2**]. We restrict to Theorem 1; similar ideas apply to proving Theorem 2. So let μ

be a finite non-negative Borel measure on $\mathbb{C}$. The first step is to transform a part of the problem to the examination of a more restricted class of measures using the tangent measures introduced by D. Preiss in [P]. This means that we blow-up μ from small discs to the unit disc and take weak limits of sequences of such blow-ups. More formally, we say that a locally finite non-zero Borel measure ν is a tangent measure of μ at a point $a \in \mathbb{C}$, and denote $\nu \in \mathrm{Tan}(\mu, a)$, if there exist sequences $r_i \downarrow 0$ and $0 < c_i < \infty$ such that $c_i T_{a,r_i} \mu$ converges weakly to μ, where

$$T_{a,r_i} \mu(A) = \mu(r_i A + a).$$

The constants c_i are needed to keep the limit non-zero and locally finite. Often one can take $c_i = \left(\mu B(a, r_i)\right)^{-1}$. It is not very hard to show, see [P, Theorem 2.5], that for any locally finite Borel measure μ such tangent measures exist at μ almost all points $a \in \mathbb{C}$.

Note that (1) means

$$\lim_{0<\varepsilon<\delta\downarrow 0} \int_{\varepsilon<|\zeta-z|\leq\delta} (\zeta - z)^{-1} \, d\mu\zeta = 0.$$

If this and (2) hold for μ almost all $z \in \mathbb{C}$, all tangent measures $\nu \in \mathrm{Tan}(\mu, a)$ at μ almost all points $a \in \mathbb{C}$ satisfy the much stronger condition:

$$\int_{r<|\zeta-z|\leq R} (\zeta - z)^{-1} \, d\nu\zeta = 0 \quad \text{for } z \in \mathrm{spt}\,\nu,\ 0 < r < R < \infty. \tag{4}$$

Thus we are led to study locally finite non-negative Borel measures ν satisfying (4). We shall call them *symmetric*. Using a simple approximation one can show that (4) is equivalent to the following perhaps more appealing condition:

$$\frac{1}{\nu B(z,r)} \int_{B(z,r)} \zeta \, d\nu\zeta = z \quad \text{for } z \in \mathrm{spt}\,\nu,\ 0 < r < \infty.$$

This means that whenever we restrict ν to a disc centered at $\mathrm{spt}\,\nu$, the center of mass of this restriction is the center of the disc. Similarly one can also show that if ν is a symmetric measure, then

$$\psi * \nu(z) = 0 \quad \text{for } z \in \mathrm{spt}\,\nu \tag{5}$$

for any function ψ of the form $\psi(z) = z\,\varphi(|z|)$, where φ is a non-negative Borel function on $\mathbb{R}$ with compact support. In fact, this holds also for φ's which decay exponentially at infinity. In order to prove this one needs first to verify the doubling condition

$$\limsup_{R\to\infty} \nu B(0, 2\,R)/\nu B(0, R) < \infty$$

by some technical work.

We start to analyze the structure of a symmetric measure ν by choosing $\psi(z) = z\,e^{-|z|^2}$ in (5). Then

$$\mathrm{spt}\,\nu \subset (\psi * \nu)^{-1}\{0\}.$$

If this zero-set is all of C, that is $\psi * \nu = 0$, we have for the Fourier transforms

$$0 = (\psi * \nu)^\wedge = \hat{\psi}\,\hat{\nu}.$$

This means that the support of the distribution $\hat{\nu}$ is contained in the zero-set of $\hat{\psi}$. But $\hat{\psi}(\zeta) = c\,\zeta\,e^{-|\zeta|^2}$, and so $\mathrm{spt}\,\hat{\nu} = \{0\}$, which is to say that $\hat{\nu}$ is a linear combination of the Dirac delta δ_0 and its partial derivatives. Then ν must be a polynomial, but a polynomial (that is, the Lebesgue measure multiplied by a polynomial) can obviously be symmetric only if it is constant.

Thus we have either $\nu = c\mathcal{L}^2$, where $\mathcal{L}^2$ is the Lebesgue measure, or $\mathrm{spt}\,\nu \subset (\psi * \nu)^{-1}\{0\} \neq C$. The function $\psi * \nu$ is real analytic, whence in the latter case $\mathrm{spt}\,\nu$ is contained in countably many smooth curves. Using this and symmetricity, one can proceed to prove the following theorem:

Theorem 4. *Let ν be a non-negative locally finite Borel measure on C.*

(1) If ν is symmetric, it is either continuous ($\nu\{a\} = 0$ for $a \in C$) or discrete ($\nu = \sum c_i \delta_{a_i}$ for some discrete set $\{a_i\}$).

(2) If ν is continuous, it is symmetric if and only if it is one of the following three alternatives:

(i) $\nu = c\mathcal{L}^2$ *for some* $0 < c < \infty$,
(ii) $\nu = c\mathcal{H}^1|L$ *for some line* L *and some* $0 < c < \infty$,
(iii) $\nu = c\sum_{m\in\mathbb{Z}}\mathcal{H}^1|(L + 2mw) + d\sum_{m\in\mathbb{Z}}\mathcal{H}^1|(L + (2m+1)w)$

for some line L, some non-zero complex number w which is as a vector orthogonal to L, and some positive numbers c and d.

Thus we know now pretty much about tangent measures for measures μ such that $\liminf_{r\downarrow 0}\mu B(z,r)/r > 0$ and $C_\mu(z)$ exists for μ almost all $z \in C$. However, this is not quite enough. But with various arguments, which I shall not describe here, we can make use of all this information to say more:

Theorem 5. *If (1) and (2) hold for μ almost all $z \in C$, then for μ almost all $a \in C$ either $\mu\{a\} > 0$ or every $\nu \in \mathrm{Tan}(\mu, a)$ is linear, that is, $\nu = c\mathcal{H}^1|L$ for some line L through the origin and some $0 < c < \infty$.*

Now the second alternative conclusion in Theorem 5 means that μ looks locally much like a nice measure on a rectifiable curve, and as a matter of fact it is sufficient when combined with (2) for the rectifiability of μ in our generalized sense. By a polygonal approximation one can prove:

Theorem 6. *If for μ almost all $z \in C$ (2) holds and every $\nu \in \mathrm{Tan}(\mu, z)$ is linear, then μ is rectifiable.*

Theorem 1 follows now immediately from Theorems 5 and 6.

I don't know whether the assumption (2) in Theorem 1 could be replaced by the weaker condition

$$\limsup_{r\downarrow 0}\mu B(z,r)/r > 0 \tag{6}$$

for μ almost all $z \in \mathbf{C}$. The method breaks down because Preiss showed in [P] that Theorem 6 is false with (2) replaced by (6). (In fact, (2) is used also elsewhere in the proof of Theorem 1, but maybe not as essentially.) Perhaps Preiss's example could be modified to show that also Theorem 1 is false with (2) replaced by (6).

References

[CCFJR] A. P. Calderón, C. P. Calderón, E. B. Fabes, M. Jodeit Jr. and N. M. Riviere, Applications of the Cauchy integrals along Lipschitz curves, Bull. Amer. Math. Soc. 84 (1978), 287–290.

[D] G. David, Opérateurs intégraux singuliers sur certaines courbes du plan, Ann. Sci. École Norm. Sup. (4) 17 (1984), 157–189.

[FK] K. J. Falconer, Geometry of Fractal Sets, Cambridge University Press, 1985.

[FX] X. Fang, The Cauchy integral of Calderón and analytic capacity, Ph. D. Thesis, Yale University, 1990.

[G] J. Garnett, Analytic Capacity and Measure, Lecture Notes in Math. 297, Springer-Verlag, 1972.

[J] P. W. Jones, Square functions, Cauchy integrals, analytic capacity, and harmonic measure, (Harmonic Analysis and Partial Differential Equations (ed. J. García-Cuerva)) , Lecture Notes in Math. 1384, Springer-Verlag, 1989, pp. 24–68.

[K] D. Khavinson, On a geometric localization of the Cauchy potentials, Michigan Math. J. 33 (1986), 377–385.

[MP1] P. Mattila, A class of sets with zero length and positive analytic capacity, Ann. Acad. Sci. Fenn. Ser. A I Math. 10 (1985), 387–395.

[MP2] P. Mattila, Cauchy singular integrals and rectifiability of measures in the plane, to appear in Adv. in Math.

[MT] T. Murai, A Real Variable Method for the Cauchy Transform, and Analytic Capacity, Lecture Notes in Math. 1307, Springer-Verlag, 1988.

[P] D. Preiss, Geometry of measures in $\mathbf{R}^n$. Distribution, rectifiability, and densities, Ann. of Math. 125 (1987), 537–643.

Added in May 1991: Recently M. S. Melnikov and the author have shown that the countable decomposition in Theorem 3 is not needed. We have proven that the finite principal values $C_\mu(z)$ exist for $\mathcal{H}^1$ almost all $z \in E$ whenever E is an $\mathcal{H}^1$ measurable rectifiable set with $\mathcal{H}^1(E) < \infty$ and μ is a finite complex Borel measure on $\mathbf{C}$.

Extension theorems for real variable
Hardy and Hardy-Sobolev spaces

Akihiko MIYACHI

1. Introduction

For $0 < p \leq 1$, let $H_p(R^n)$ denote the real variable Hardy space
as given in the paper of C. Fefferman and E. M. Stein [2; Section 11].
For $0 < p \leq 1$ and $k \in N$ (= the set of positive integers), we define
$H_p^k(R^n)$ as the set of the distributions f on R^n for which the de-
rivatives $\partial^\alpha f$ belong to $H_p(R^n)$ for $|\alpha| = k$. (Contrary to the cus-
tom, we do not require $\partial^\alpha f \in H_p(R^n)$ for $|\alpha| < k$.) We shall consider
$H_p^k(R^n)$ only for $k > n/p-n$.

Let Ω be an open subset of R^n. We denote by $H_p(R^n)|\Omega$
($H_p^k(R^n)|\Omega$ resp.) the set of the restrictions to Ω of the members
of $H_p(R^n)$ ($H_p^k(R^n)$ resp.). On the other hand, following [5], we can
directly define $H_p(\Omega)$ and $\ulcorner H_p^k(\Omega)$ (see Sections 2 and 3 of the present
paper). These spaces satisfy $H_p(\Omega) \supset H_p(R^n)|\Omega$ and $H_p^k(\Omega) \supset H_p^k(R^n)|\Omega$.
The purpose of this paper is to study the reverse inclusion, or, more
generally, to consider the possibility of extending $f \in H_p(\Omega)$ ($H_p^k(\Omega)$
resp.) to an element of $H_p(R^n)$ ($H_p^k(R^n)$ resp.) for each f individ-
ually.

The main results of this paper are as follows. We show that we
can associate each $f \in H_p^k(\Omega)$, for arbitrary Ω and for $k > n/p-n$,
with a smooth function f_1 on Ω in such a way that $f-f_1$ belongs
to $H_p^k(R^n)|\Omega$ (Theorem 3). Thus the problem of extending $f \in H_p^k(\Omega)$
to an element of $H_p^k(R^n)$ reduces to the problem of extending f_1. This
latter problem is easier to handle since the function f_1 is gentler
than f; the behavior of f_1 is gentle "inside" Ω but it inherits
that of f near the boundary of Ω. Using this result, we prove that
if Ω satisfies P. Jones's condition (see [3] or Section 3 of the pres-
ent paper) and if $k > n/p-n$ then $H_p^k(\Omega) = H_p^k(R^n)|\Omega$ (Theorem 4).
Similar results for $H_p(\Omega)$ are essentially contained in [5]; we give
them in a form slightly different from [5] (Theorems 1 and 2).

Throughout this paper, we use the letters Ω and C in the following manner. The letter Ω denotes an open subset of R^n. Unless
further restrictions are explicitly stated, Ω denotes an arbitrary
open subset of R^n. The letter C denotes positive constant which
need not be the same in different occurrences. The constant C depends

only on the dimension n of the Euclidean space considered and other explicitly indicated parameters.

Notation. By a cube we mean a subset of R^n of the form $Q = \{(x_i) \in R^n \mid a_i < x_i \le a_i + t \; (i = 1, \ldots, n)\}$, where t is called the side length and is denoted by $\ell(Q)$. If $a_i = 2^k m_i$ and $t = 2^k$ with k and m_i integers, then the corresponding cube Q is called a dyadic cube. If Q is a cube and $b > 0$, then bQ denotes the cube with the same center as Q and with side length b times as large as Q. The Lebesgue measure of a measurable subset $E \subset R^n$ is denoted by $|E|$. If E is a measurable subset of R^n and f is a measurable function on E, then we write

$$\int_E f = \int_E f(x) dx,$$

$$\|f\|_{p,E} = \left(\int_E |f|^p \right)^{1/p} \quad \text{for} \quad 0 < p < \infty,$$

and

$$\|f\|_{\infty,E} = \text{ess. sup. of} \quad |f| \quad \text{on} \quad E.$$

If $E = R^n$, then $\|\cdot\|_{p,E}$ and $\int_E f$ are simply written as $\|\cdot\|_p$ and $\int f$ respectively. The symbol Π_k, $k \in N \cup \{0\}$, denotes the set of polynomial functions on R^n of degree not exceeding k. We shall introduce other notations as we go along.

2. H_p space on Euclidean domain

We begin by recalling a result of [4].
Take a function $\phi \in C_0^\infty(R^n)$ such that $\text{supp } \phi \subset \{x \mid |x| < 1\}$ and $\int \phi = 1$. For $t > 0$, set $(\phi)_t(x) = t^{-n}\phi(t^{-1}x)$. For $f \in D'(\Omega)$ (= the set of distributions on Ω), we define

$$f_{\phi,\Omega}^+(x) = \sup\{ |\langle f, (\phi)_t(x-\cdot)\rangle| \mid 0 < t < \text{dis}(x, \Omega^c)\}, \quad x \in \Omega.$$

Here $\text{dis}(\cdot, \cdot)$ denotes Euclidean distance, and $\langle f, \psi \rangle$ with a distribution f and a testing function ψ denotes $f(\psi)$.
For $f \in D'(\Omega)$ and $k \in N$, we define

$$f_{k,\Omega}^*(x) = \sup\{ |\langle f, \psi \rangle| \}, \quad x \in R^n,$$

where the supremum is taken over those $\psi \in C_0^\infty(R^n)$ for which there exist $t = t_\psi > 0$ such that $\text{supp } \psi \subset \{y \mid |y-x| < t\} \cap \Omega$ and $\|\partial^\alpha \psi\|_\infty \le t^{-n-|\alpha|}$ for $|\alpha| \le k$.
The following theorem follows from the result of [4].

Theorem A. Let $f \in D'(\Omega)$ and let ϕ be a function as mensioned above. If $k \in N$ and $n/(n+k) < p \le \infty$, then

$$\| f^*_{k,\Omega} \|_p \le C \| f^+_{\phi,\Omega} \|_{p,\Omega} \ ,$$

where the constant C depends only on n, k, p, and ϕ.

Following [5], we define $H_p(\Omega)$, $0 < p \le 1$, as the set of those $f \in D'(\Omega)$ for which $f^+_{\phi,\Omega}$ belong to $L_p(\Omega)$. By Theorem A, $H_p(\Omega)$ does not depend on ϕ.

For proper open subset Ω of R^n, we define $G(\Omega)$ as the set of maximal dyadic cubes Q satisfying $4Q \subset \Omega$.

Lemma 1. If Ω is a proper open subset of R^n, then $G(\Omega)$ has the following properties.

(i) $G(\Omega)$ is a disjoint family and the union of all the cubes in $G(\Omega)$ is equal to Ω.

(ii) If $Q \in G(\Omega)$ and $x \in 3Q$, then $C^{-1} \le \ell(Q)/dis(x,\Omega^c) \le C$.

(iii) If $Q, Q' \in G(\Omega)$ and the closures $\overline{Q}$ and $\overline{Q}'$ have nonempty intersection, then $1/2 \le \ell(Q)/\ell(Q') \le 2$.

(iv) The overlap of the family $\{3Q| \ Q \in G(\Omega)\}$ does not exceed C, i. e., no point of R^n belong to more than C of the cubes $3Q$, $Q \in G(\Omega)$.

(v) There exist functions ϕ^Ω_Q, $Q \in G(\Omega)$, such that $\phi^\Omega_Q \in C^\infty_0(R^n)$, supp $\phi^\Omega_Q \subset 2Q$, $0 \le \phi^\Omega_Q(x) \le 1$, $\| \partial^\alpha \phi^\Omega_Q \|_\infty \le C_\alpha \ell(Q)^{-|\alpha|}$ for each α, and $\sum_{Q \in G(\Omega)} \phi^\Omega_Q(x) = 1$ for all $x \in \Omega$.

For the proof of this lemma, see, e. g., [7; Chapter VI, Section 1].

Now, let Ω be a proper open subset of R^n, $0 < p \le 1$, and $f \in H_p(\Omega)$. For each $Q \in G(\Omega)$, let P_Q be the polynomial in $\Pi_{[n/p-n]}$ such that

$$\langle f\phi^\Omega_Q - P_Q \chi_Q \ , \ P \rangle = 0 \quad \text{for all} \quad P \in \Pi_{[n/p-n]} \ .$$

Here χ_Q denotes the characteristic function of Q. We set

$$f_0 = \sum_{Q \in G(\Omega)} P_Q \chi_Q \ .$$

Then we have the following theorem.

Theorem 1. Let Ω, p, f, and f_0 be as above. Then $f - f_0$ belongs to $H_p(R^n)|\Omega$.

<u>Proof</u>. Set $k = [n/p-n]+1$ and $f^Q = f\phi_Q^\Omega - P_Q\chi_Q$. Note that f^Q is a compactly supported distribution on R^n and supp $f^Q \subset 2Q$.

We first regard f^Q as a distribution on $3Q°$, where $3Q°$ denotes the interior of $3Q$. Then, by the result of [5; Section 3], we have the decomposition

$$(2.1) \qquad f^Q = \sum_i \lambda_i^Q g_i^Q + \sum_j \mu_j^Q h_j^Q \quad \text{on} \quad 3Q° ,$$

where g_i^Q are the usual p-atoms, h_j^Q are $(p,3Q°)$-atoms (see [5; p.210]), λ_i^Q and μ_j^Q are nonnegative numbers such that

$$(2.2) \qquad \sum_i (\lambda_i^Q)^p + \sum_j (\mu_j^Q)^p \leq C_p \int_{3Q°} ((f^Q)_{k,3Q°}^*)^p .$$

Since supp $f^Q \subset 2Q$ and $\langle f^Q, P \rangle = 0$ for all $P \in \Pi_{k-1}$, the argument in [5; Section 3] gives the following further information on the decomposition (2.1). Firstly, there exists a compact subset of $3Q°$ which includes all the supports of g_i^Q and h_j^Q . This implies that both sides of (2.1) are compactly supported distributions on R^n and the equality (2.1) holds as distributions on R^n . Secondly, the number of the nonzero h_j^Q appearing in (2.1) does not exceed C and $\|h_j^Q\|_\infty$ $\leq C_p|Q|^{-1/p}$. Thirdly, $\langle \sum_j \mu_j^Q h_j^Q, P \rangle = 0$ for all $P \in \Pi_{k-1}$. From these facts, we can rewrite (2.1)-(2.2) as

$$f^Q = \sum_j \nu_j^Q f_j^Q \quad \text{on} \quad R^n ,$$

where f_j^Q are p-atoms and ν_j^Q are nonnegative numbers such that

$$\sum_j (\nu_j^Q)^p \leq C_p \int_{3Q°} ((f^Q)_{k,3Q°}^*)^p .$$

On the other hand, it is easy to see that

$$(f^Q)_{k,3Q°}^*(x) \leq C_p f_{k,\Omega}^*(x), \quad x \in 3Q° .$$

Hence, using Lemma 1, (iv), we obtain

$$\sum_Q \sum_j (\nu_j^Q)^p \leq C_p \sum_Q \int_{3Q°} (f_{k,\Omega}^*)^p \leq C_p \int_\Omega (f_{k,\Omega}^*)^p < \infty ,$$

where the summation on Q is taken over $Q \in G(\Omega)$. Hence the sum

$$\sum_Q f^Q = \sum_Q \sum_j \nu_j^Q f_j^Q$$

converges in $H_p(R^n)$. Since $f-f_0 = \sum_Q f^Q$ on Ω , we have $f-f_0 \in$ $H_p(R^n)|\Omega$. Theorem 1 is proved.

Remark. (1) The above proof showed that $f-f_0$ can be written as

(2.3)
$$f-f_0 = \sum_{j=1}^{\infty} \lambda_j g_j \quad \text{on} \quad \Omega$$

with p-atoms g_j whose supports are included in Ω and with $\lambda_j \geq 0$ satisfying $\sum_j (\lambda_j)^p < \infty$.

(ii) By examining the argument in the above proof and in [5; Section 3], we can prove the following: If $f \in H_p(\Omega) \cap L_1(\Omega)$ with $0 < p \leq 1$, then $f_0 \in L_1(\Omega)$ and we can rearrange the series $\sum_j \lambda_j g_j$ in (2.3) so that some subsequence of the partial sum sequence $\{S_m\}$, $S_m = \sum_{j=1}^{m} \lambda_j g_j$, converges in $L_1(\Omega)$.

We shall give a simple application of Theorem 1 and the above Remark.

If $x = (x_1, \ldots, x_n) \in R^n$, we write $x' = (x_1, \ldots, x_{n-1}) \in R^{n-1}$. Let h be a real-valued Lipschitz function on R^{n-1} and let

$$\Omega_h = \{x \in R^n |\ x_n > h(x')\}.$$

The result of [5; Section 6] shows that $H_p(\Omega_h) = H_p(R^n)|\Omega_h$ for all $p \in (0,1]$ and that there exists an extension operator $H_p(\Omega_h) \to H_p(R^n)$. Here we shall prove that the odd reflection works as an extension operator for p near 1. The odd reflection E is the bounded operator $L_1(\Omega_h) \to L_1(R^n)$ which is defined by

$$Ef(x) = \begin{cases} f(x) & \text{if } x \in \Omega_h \\ -f(x',2h(x')-x_n) & \text{if } x_n < h(x'). \end{cases}$$

The precise statement is in the following theorem. The case $p = 1$ of this theorem is due to Strichartz [8].

Theorem 2. Let h and Ω_h be as above, and let $f \in L_1(\Omega_h)$ and $n/(n+1) < p \leq 1$. Then $f \in H_p(R^n)|\Omega_h$ if and only if $Ef \in H_p(R^n)$.

Proof. The "if" part is obvious. We shall prove the "only if" part. We write $\Omega = \Omega_h$ and $G = G(\Omega_h)$. Suppose $f \in L_1(\Omega) \cap H_p(R^n)|\Omega$. Then $f \in H_p(\Omega)$. It is easy to see that if g is a p-atom with support included in Ω then $Eg-g$ is a constant multiple of a p-atom. (Note that, since $n/p-n < 1$, the only moment condition on p-atom is vanishing of its integral.) Hence, by the above Remark, we see that $E(f-f_0) = \sum_j \lambda_j Eg_j$ belongs to $H_p(R^n)$. On the other hand, we have

$$Ef_0 = \sum_{Q \in G} P_Q Ex_Q$$

with $|P_Q| \leq C \inf\{f_{1,\Omega}^*(x) | x \in Q\}$. (Note that P_Q in the definition of f_0 is now a constant function.) It is easy to see that $|Q|^{-1/p} E\chi_Q$ for $Q \in G$ is a constant multiple of a p-atom. We also have

$$\sum_{Q \in G} |P_Q|^p |Q| \leq \sum_{Q \in G} C_p \int_Q (f_{1,\Omega}^*)^p = C_p \int_\Omega (f_{1,\Omega}^*)^p < \infty.$$

Hence $Ef_0 \in H_p(R^n)$. Thus $Ef = E(f-f_0) + Ef_0 \in H_p(R^n)$. Theorem 2 is proved.

3. H_p^k space on Euclidean domain

For $0 < p \leq 1$ and $k \in N$, we define $H_p^k(\Omega)$ as the set of those $f \in D'(\Omega)$ for which $\partial^\alpha f \in H_p(\Omega)$ for $|\alpha| = k$.

For measurable functions f on Ω and for $k \in N$ and $0 < q < \infty$, we define $f_{k,q}^b(x)$, $x \in \Omega$, by

$$f_{k,q}^b(x) = \sup_{\substack{Q:cube \\ x \in Q \subset \Omega}} \inf_{P \in \Pi_{k-1}} |Q|^{-k/n - 1/q} \| f - P \|_{q,Q}.$$

This maximal function was introduced by DeVore and Sharpley [1].

The following lemma will be used afterwards. For the proof of this lemma, cf. [1; Theorem 4.3].

Lemma 2. If $k \in N$ and $0 < q < p < \infty$, then

$$\| f_{k,q}^b \|_{p,\Omega} \leq \| f_{k,p}^b \|_{p,\Omega} \leq C_{k,p,q} \| f_{k,q}^b \|_{p,\Omega}.$$

The next theorem is given in [6].

Theorem B. Let $k \in N$, $0 < p \leq 1$, and $k > n/p-n$. Then:

(i) If $f \in H_p^k(\Omega)$, then $f \in L_{1,loc}(\Omega)$ and $f_{k,p}^b \in L_p(\Omega)$;

(ii) Conversely, if f is a measurable function on Ω and $f_{k,p}^b \in L_p(\Omega)$, then $f \in L_{1,loc}(\Omega)$ and $f \in H_p^k(\Omega)$.

By virtue of Theorem B and Lemma 2, if k and p satisfy the assumption of Theorem B and if $0 < q \leq p$, then $H_p^k(\Omega)$ can be identified with the set of measurable functions f on Ω such that $f_{k,q}^b \in L_p(\Omega)$. From now on we shall freely use this identification.

Suppose Ω is a proper open subset of R^n and k and p satisfy the assumption of Theorem B, and let $f \in H_p^k(\Omega)$. Take a number q such that $0 < q < p$ and $k > n/q-n$. For each cube Q in $G(\Omega)$, take a polynomial π_Q in Π_{k-1} such that

$$\| f - \pi_Q \|_{q,Q} = \min\{ \| f - P \|_{q,Q} | \ P \in \Pi_{k-1} \}.$$

We set

$$f_1(x) = \sum_{Q \in G(\Omega)} \pi_Q(x) \phi_Q^{\Omega}(x), \quad x \in \Omega.$$

Theorem 3. Let Ω, k, p, f, and f_1 be as mensioned above. Then $f - f_1 \in H_p^k(R^n)|\Omega$. Moreover, the function

$$(f - f_1)^{\sim}(x) = \begin{cases} f(x) - f_1(x) & \text{if} \ \ x \in \Omega \\ 0 & \text{if} \ \ x \in \Omega^c \end{cases}$$

belongs to $H_p^k(R^n)$.

In fact, by almost the same argument as in [6; Section 5, Proof of Theorem 3], we can prove that $(f - f_1)^{\sim}$ is a locally integrable function on R^n and that the pointwise inequality

$$(3.1) \qquad \left(\partial^{\alpha}((f - f_1)^{\sim}) \right)^*_{k,R^n}(x) \le C_{k,q} \ M_q((f_{k,q}^b)^{\sim})(x), \quad x \in R^n,$$

holds for $|\alpha| = k$, where M_q is defined by

$$M_q(g)(x) = \sup_{\substack{Q:\text{cube} \\ Q \ni x}} \left(\frac{1}{|Q|} \int_Q |g|^q \right)^{1/q}, \quad x \in R^n,$$

and $(f_{k,q}^b)^{\sim}$ denotes the extension of $f_{k,q}^b$ to R^n that is equal to 0 outside Ω. Since $(f_{k,q}^b)^{\sim} \in L_p(R^n)$ and M_q maps $L_p(R^n)$ into itself (since $q < p$), (3.1) implies that $\partial^{\alpha}((f - f_1)^{\sim}) \in H_p(R^n)$ for $|\alpha| = k$ or equivalently $(f - f_1)^{\sim} \in H_p^k(R^n)$. Details of the proof of (3.1) are left to the reader.

Next we shall give a sufficient condition on Ω to satisfy $H_p^k(\Omega) = H_p^k(R^n)|\Omega$. Before we state the result, we shall recall Jones's condition (cf. [3]).

Let Ω be a connected proper open subset of R^n. For $x \in \Omega$, we set $\delta(x) = \text{dis}(x,\Omega^c)$. For $x, y \in \Omega$, we define

$$d_1(x,y) = \inf\{ \int_{\gamma} \frac{|dz|}{\delta(z)} \},$$

where the infimum is taken over all rectifiable curves γ in Ω that join x to y. Also for $x, y \in \Omega$, we define

$$d_2(x,y) = \left| \log \frac{\delta(x)}{\delta(y)} \right| + \log\left(2 + \frac{\text{dis}(x,y)}{\delta(x) + \delta(y)} \right).$$

We say that Ω satisfies Jones's condition if there exists a constant $K > 0$ such that $d_1(x,y) \leq K d_2(x,y)$ for all x, $y \in \Omega$. It is not difficult to show that Jones's condition is equivalent to the following condition: For any two cubes Q, $Q' \in G(\Omega)$, there exist cubes $Q_j \in G(\Omega)$, $j = 0, 1, \ldots, m$, such that $Q_0 = Q$, $Q_m = Q'$, $\overline{Q}_j \cap \overline{Q}_{j-1} \neq \emptyset$ ($j = 1, \ldots, m$) and

$$m \leq K' \left[\left| \log \frac{\ell(Q)}{\ell(Q')} \right| + \log\left(2 + \frac{dis(Q,Q')}{\ell(Q)+\ell(Q')}\right) \right],$$

where K' is a constant independent of Q and Q'.

$\underline{\text{Theorem}}$ 4. If Ω is a connected proper open subset of R^n satisfying Jones's condition, and if k and p satisfy the assumption of Theorem B, then $H_p^k(\Omega) = H_p^k(R^n)|\Omega$.

$\underline{\text{Proof}}$. The inclusion $H_p^k(R^n)|\Omega \subset H_p^k(\Omega)$ holds for arbitrary Ω and for all $k \in N$ and all $p \in (0,1]$. We shall prove the reverse inclusion for Ω, k, and p satisfying the assumptions of the theorem. We may assume Ω is not empty.

Let $\hat{\Omega}$ denote the interior of Ω^c. We write $G = G(\Omega)$, $\hat{G} = G(\hat{\Omega})$, and $\phi_Q = \phi_Q^\Omega$ if $Q \in G$ and $\phi_Q = \phi_Q^{\hat{\Omega}}$ if $Q \in \hat{G}$. Set $L = \sup \{\ell(Q)|$ $Q \in G\}$ ($\leq \infty$). For each $Q \in \hat{G}$, we take $\check{Q} \in G$ as follows. If $\ell(Q) \leq L$, then $\check{Q} \in G$ is a cube which is nearest to Q in the Euclidean metric satisfying $\ell(\check{Q}) = \ell(Q)$. (By Lemma 1, (iii), there exists such $\check{Q}$.) If $L < \infty$, then we take a cube $Q_* \in G$ with $\ell(Q_*) = L$ and set $\check{Q} = Q_*$ for all those $Q \in \hat{G}$ with $\ell(Q) > L$.

The following are consequences of Jones's condition.
(†) If $Q \in \hat{G}$ and $\ell(Q) \leq L$, then $dis(Q,\check{Q}) \leq C_K \ell(Q)$.
(††) If $L < \infty$, then $\Omega \subset \{x | dis(x,Q_*) < C_K L\}$.
(Here K is the constant in Jones's condition.) See [3; Lemma 2.10 and p. 57].

Now suppose $f \in H_p^k(\Omega)$. Let q, π_Q ($Q \in G$), and f_1 be the same as in Theorem 3. By Theorem 3, it is sufficient to prove that $f_1 \in H_p^k(R^n)|\Omega$. For each $Q \in \hat{G}$, we set $\pi_Q = \pi_{\check{Q}}$. We define

$$F(x) = \sum_{Q \in G \cup \hat{G}} \pi_Q(x)\phi_Q(x), \quad x \in R^n.$$

Obviously F is an extension of f_1. Hence it is sufficient to show $F \in H_p^k(R^n)$. In order to prove this, we shall prove the pointwise inequality

(3.2) $\qquad F_{k,q}^b(x) \leq C_{K,k,q} M_q\left(M_q((f_{k,q}^b)^\sim)\right)(x), \quad x \in R^n.$

Once this is proved, then $F^b_{k,q} \in L_p(R^n)$ and hence $F \in H^k_p(R^n)$ (by Lemma 2 and Theorem B) and thus the proof of the theorem will be complete. Hereafter we shall simply write f^b for $(f^b_{k,q})^\sim$.

In order to prove (3.2), we use the following estimate:

(*) $\Big($ If $Q, Q' \in G \cup \hat{G}$, $A > 0$, $\ell(Q) \le \ell(Q')$, and $\text{dis}(Q,Q') \le A\ell(Q')$,

then

(3.3) $\qquad \|\pi_Q - \pi_{Q'}\|_{\infty,Q} \le C_{K,k,q,A,B}\, \ell(Q')^k \inf_{BQ} M_q(f^b)$

for each $B \ge 1$.

We shall prove this estimate in several steps.

(1°) If $Q, Q' \in G$ and $\overline{Q} \cap \overline{Q}' \ne \emptyset$, then

$$\|\pi_Q - \pi_{Q'}\|_{\infty,Q} \le C_{k,q}\, |Q|^{k/n - 1/q} \Big(\int_{Q \cup Q'} (f^b)^q\Big)^{1/q}.$$

As for this estimate, cf. [1; Section 4, (4.9), (4.9)'].

(2°) If $Q, Q' \in G$, $A > 0$, $1/2 \le \ell(Q)/\ell(Q') \le 2$, and $\text{dis}(Q,Q') \le A\ell(Q)$, then (3.3) holds. In fact, by Jones's condition, there exist cubes $Q_j \in G$ ($j = 0, 1, \ldots, m$) such that $Q_0 = Q$, $Q_m = Q'$, $\overline{Q}_j \cap \overline{Q}_{j-1} \ne \emptyset$, and $m \le C_{K,A}$. We have $1/C_{K,A} \le \ell(Q_j)/\ell(Q) \le C_{K,A}$, $Q \subset C_{K,A}Q_j$, and $Q_j \subset C_{K,A}Q$. By (1°),

(3.4) $\quad \|\pi_{Q_j} - \pi_{Q_{j-1}}\|_{\infty,Q_j} \le C_{k,q}\, |Q_j|^{k/n - 1/q} \Big(\int_{Q_j \cup Q_{j-1}} (f^b)^q\Big)^{1/q}.$

Since $C_{K,A}Q_j \supset Q$, the left hand side of (3.4) is not less than

$$\|\pi_{Q_j} - \pi_{Q_{j-1}}\|_{\infty,Q} / C_{K,A,k}.$$

On the other hand, since $\ell(Q_j) \approx \ell(Q)$ as above and $Q_j \subset C_{K,A}Q$, the right hand side of (3.4) is not larger than

$$C_{K,k,q,A}\, |Q|^{k/n - 1/q} \Big(\int_{C_{K,A}Q} (f^b)^q\Big)^{1/q}$$

$$\le C_{K,k,q,A,B}\, \ell(Q')^k \inf_{BQ} M_q(f^b), \qquad B \ge 1.$$

Hence, for $B \ge 1$,

$$\|\pi_{Q_j} - \pi_{Q_{j-1}}\|_{\infty,Q} \le C_{K,k,q,A,B}\, \ell(Q')^k \inf_{BQ} M_q(f^b).$$

Summing up the last inequalities for $j = 1, \ldots, m$ with $m \le C_{K,A}$, we obtain (3.3).

(3°) Proof of (*) in the case $Q, Q' \in G$. By Jones's condition,

we can find cubes $Q_j \in G$ ($j = 0, 1, \ldots, m$) such that $Q_0 = Q$, $\ell(Q_j)$
$= 2\ell(Q_{j-1})$, $\ell(Q_m) = \ell(Q')$, $\mathrm{dis}(Q_j, Q_{j-1}) \leq C_K \ell(Q_j)$, $\mathrm{dis}(Q, Q_j) \leq$
$C_K \ell(Q_j)$, and $\mathrm{dis}(Q_m, Q') \leq C_{K,A} \ell(Q')$ (see [3; Lemmas 2.6 and 2.7]).
By ($2°$),

$$\| \pi_{Q_j} - \pi_{Q_{j-1}} \|_{\infty, Q_j} \leq C_{K,k,q,B} \, \ell(Q_j)^k \, \inf_{BQ_j} M_q(f^b)$$

and

$$\| \pi_{Q_m} - \pi_{Q'} \|_{\infty, Q_m} \leq C_{K,k,q,A,B} \, \ell(Q')^k \, \inf_{BQ_m} M_q(f^b)$$

for each $B \geq 1$. Since $C_K Q_j \supset Q$, the above estimates imply

(3.5) $$\| \pi_{Q_j} - \pi_{Q_{j-1}} \|_{\infty, Q} \leq C_{K,k,q,B} \, \ell(Q_j)^k \, \inf_{BQ} M_q(f^b)$$

and

(3.6) $$\| \pi_{Q_m} - \pi_{Q'} \|_{\infty, Q} \leq C_{K,k,q,A,B} \, \ell(Q')^k \, \inf_{BQ} M_q(f^b).$$

Note that

$$\sum_{j=1}^{m} \ell(Q_j)^k = \sum_{j=1}^{m} (2^{j-m} \ell(Q'))^k \leq C_k \ell(Q')^k.$$

Hence, summing up (3.5) for $j = 1, \ldots, m$ and (3.6), we obtain (3.3).

($4°$) Proof of ($*$) in the case $Q \in G$ and $Q' \in \hat{G}$. In this case,
using ($\dagger$) and ($\dagger\dagger$), we see that $\check{Q}' \in G$, $\ell(Q) \leq \ell(\check{Q}') \leq \ell(Q')$, and
$\mathrm{dis}(Q, \check{Q}') \leq C_{K,A} \ell(\check{Q}')$. Hence the proof reduces to ($3°$).

($5°$) Proof of ($*$) in the case $Q \in \hat{G}$ and $Q' \in G$. In this case,
$\check{Q} \in G$, $\ell(\check{Q}) = \ell(Q) \leq \ell(Q')$, and $\mathrm{dis}(\check{Q}, Q') \leq C_{K,A} \ell(Q')$ (by ($\dagger$)).
Hence, by ($3°$), we have

$$\| \pi_{\check{Q}} - \pi_{Q'} \|_{\infty, \check{Q}} \leq C_{K,k,q,A,B} \, \ell(Q')^k \, \inf_{B\check{Q}} M_q(f^b)$$

for each $B \geq 1$. Since $\pi_{\check{Q}} = \pi_Q$ and $C_K \check{Q} \supset Q$ (by ($\dagger$)), the above
estimate implies (3.3).

($6°$) Proof of ($*$) in the case $Q, Q' \in \hat{G}$. If $\ell(Q) > L$, then π_Q
$= \pi_{Q'} = \pi_{Q_*}$ and hence (3.3) is trivial. If $\ell(Q) \leq L$, then, using
($\dagger$) and ($\dagger\dagger$), we see that $\check{Q}' \in G$, $\ell(Q) \leq \ell(\check{Q}') \leq \ell(Q')$, and
$\mathrm{dis}(Q, \check{Q}') \leq C_{K,A} \ell(\check{Q}')$. Hence the proof reduces to ($5°$).

Thus we have proved ($*$).

Now we shall prove (3.2). Fix an $x \in R^n$ and a (not necessarily dyadic) cube S containing x. We shall prove that there exists a polynomial P in Π_{k-1} such that

$$(3.7) \qquad |S|^{-k/n-1/q} \, \|F-P\|_{q,S} \leq C_{K,k,q} \, M_q(M_q(f^b))(x),$$

which will imply (3.2).

Case I. $2S \not\subset \Omega$ and $2S \not\subset \hat{\Omega}$. Take a cube $Q_0 \in G \cup \hat{G}$ of largest side length satisfying $2Q_0 \cap S \neq \emptyset$. As a consequence of Jones's condition, $\partial\Omega = R^n \setminus (\Omega \cup \hat{\Omega})$ has Lebesgue measure 0 (see [3; Cor. 2.9]). Hence

$$\|F-\pi_{Q_0}\|^q_{q,S} = \int_S |\sum_Q (\pi_Q - \pi_{Q_0})\phi_Q|^q = (3.8), \quad \text{say.}$$

(Here and hereafter summation over Q is taken over all $Q \in G \cup \hat{G}$ unless further restrictions are indicated.) Since $q < 1$,

$$(3.8) \leq \int_S \sum_Q |\pi_Q - \pi_{Q_0}|^q (\phi_Q)^q$$

$$\leq C \sum_{2Q \cap S \neq \emptyset} \|\pi_Q - \pi_{Q_0}\|^q_{\infty,2Q} \, |Q| = (3.9), \quad \text{say.}$$

From the assumption on S, it follows that $Q \in G \cup \hat{G}$ with $2Q \cap S \neq \emptyset$ satisfy $\ell(Q) \leq C\ell(S)$ and $Q \subset CS$. On the other hand, $\ell(Q_0) \geq \ell(S)/C_K$, which is another consequence of Jones's condition (see [3; Lemma 2.8]). Hence, if $Q \in G \cup \hat{G}$ and $2Q \cap S \neq \emptyset$, then $\ell(Q) \leq \ell(Q_0)$ (since $\ell(Q_0)$ is maximum) and

$$\text{dis}(Q,Q_0) \leq \text{diam } 2Q + \text{diam } S + \text{diam } 2Q_0 \leq C_K\ell(Q_0).$$

Thus, using (*), we have

$$(3.9) \leq C_{K,k,q} \sum_{2Q \cap S \neq \emptyset} (\ell(Q_0)^k \inf_Q M_q(f^b))^q \, |Q|$$

$$\leq C_{K,k,q} \, \ell(S)^{kq} \sum_{2Q \cap S \neq \emptyset} \int_Q (M_q(f^b))^q$$

$$\leq C_{K,k,q} \, \ell(S)^{kq} \int_{CS} (M_q(f^b))^q$$

$$\leq C_{K,k,q} \, \ell(S)^{kq+n} \, [M_q(M_q(f^b))(x)]^q.$$

Combining the above inequalities, we obtain (3.7) with $P = \pi_{Q_0}$.

Case II. $2S \subset \Omega$. Take a cube $Q_0 \in G$ of largest side length satisfying $2Q_0 \cap S \neq \emptyset$. Then the following hold for all those $Q \in G$ with $2Q \cap S \neq \emptyset$:

$$\begin{cases} \ell(Q_0)/C \leq \ell(Q) \leq \ell(Q_0), \\[2mm] C\ell(Q) \geq \ell(S), \quad CQ \supset S, \\[2mm] \text{dis}(Q,Q_0) \leq C\ell(Q_0), \end{cases}$$

(3.10)

and, for all α,

(3.11) $\qquad \|\partial^\alpha(\pi_Q - \pi_{Q_0})\|_{\infty,2Q} \leq C_{K,k,q} \ \ell(Q_0)^{k-|\alpha|} \ \inf_S M_q(f^b).$

In fact, we can easily derive (3.10) from the geometry of the situation, and we can prove (3.11) using (3.10), (*), and the inequality

$$\|\partial^\alpha P\|_{\infty,2Q} \leq C_k \ell(Q)^{-|\alpha|} \|P\|_{\infty,Q} \qquad \text{for} \quad P \in \Pi_{k-1}.$$

Now, using (3.10), (3.11), and Lemma 1, (iv) and (v), we obtain the following estimate for $|\alpha| = k$ and for all $y \in S$:

$$|\partial_y^\alpha F(y)| = |\partial_y^\alpha(F-\pi_{Q_0})(y)|$$

$$= \left| \sum_{\substack{Q \in G \\ 2Q \cap S \neq \emptyset}} \partial_y^\alpha ((\pi_Q - \pi_{Q_0})\phi_Q)(y) \right| \leq C_{K,k,q} \ \inf_S M_q(f^b).$$

Hence, if $P \in \Pi_{k-1}$ denotes the Taylor polynomial of F about x ($\in S$), then

$$\|F-P\|_{\infty,S} \leq C_{K,k,q} \ \ell(S)^k \ \inf_S M_q(f^b).$$

This estimate a fortiori implies (3.7).

     Case III. $2S \subset \hat{\Omega}$. The proof is almost the same as in Case II; infact, the only necessary change is to replace G by $\hat{G}$.

     Thus we have proved (3.2). This completes the proof of Theorem 4.

References

[1] R. A. DeVore and R. C. Sharpley, Maximal Functions Measuring Smoothness, Mem. Amer. Math. Soc., Vol. 47, No. 293, 1984.
[2] C. Fefferman and E. M. Stein, H^p spaces of several variables, Acta Math. 129(1972), 137-193.
[3] P. W. Jones, Extension theorems for BMO, Indiana Univ. Math. J. 29 (1980), 41-66.
[4] A. Miyachi, Maximal functions for distributions on open sets, Hitotsubashi J. Arts Sci. 28(1987), 45-58.
[5] A. Miyachi, H^p spaces over open subsets of R^n, Studia Math. 95(1990), 205-228.
[6] A. Miyachi, Hardy-Sobolev spaces and maximal functions, J. Math. Soc. Japan 42(1990), 73-90.
[7] E. M. Stein, Singular Integrals and Differentiability Properties of Functions, Princeton Univ. Press, 1970.

[8] R. S. Strichartz, The Hardy space H^1 on manifolds and submanifolds, Can. J. Math. 24(1972), 915-925.

Department of Mathematics, Hitotsubashi University
Kunitachi, Tokyo, 186 Japan

Boundedness of some classical operators
on generalized Morrey spaces

Takahiro Mizuhara

Department of Mathematics
Yamagata University
Yamagata 990, Japan

1. Introduction.

Let $\Phi = \Phi(r), r > 0$, be a growth function, that is, a positive increasing function in $(0, \infty)$. Let us assume that the growth function Φ satisfies a doubling condition with a doubling constant $D = D(\Phi) \geq 1$, that is,

$$\Phi(2r) \leq D\Phi(r) \quad \text{for all } r > 0 \tag{1}$$

Definition 1.1. *Let $1 \leq p < \infty$. We denote by $L^{p,\Phi} = L^{p,\Phi}(R^n)$ the space of locally integrable functions f, defined on the Euclidean n-space R^n, such that for every $x_0 \in R^n$ and every $r > 0$*

$$\int_{B(x_0,r)} |f(x)|^p dx \leq C^p \Phi(r) \tag{2}$$

where $B(x_0, r)$ is the ball centered at $x_0 \in R^n$ and with radius $r > 0$ and C is a constant depending only on f. Let $\|f\|_{p,\Phi}$ be the smallest constant C satisfying (2). Then $L^{p,\Phi}$ becomes a Banach space with norm $\| \cdot \|_{p,\Phi}$.

When $\Phi(r) = r^\lambda, \lambda \geq 0$, $L^{p,\Phi}$ is denoted simply by $L^{p,\lambda}$. $L^{p,\lambda}, 0 < \lambda < n$, is the classical Morrey spaces introduced in [10] to studying the local behaviour of solutions to second order elliptic partial differential equations and investigated by D.R.Adams [1] related to Riesz potentials, and recently also by F.Chiarenza and M.Frasca [3] in studying the boundedness of Hardy-Littlewood maximal operators, Riesz potentials and some singular integral operators. Let $1 \leq p < \infty$. Then $L^{p,0} = L^p(R^n)$ and $L^{p,n} = L^\infty(R^n)$ isometrically (see R.L.Long [9]). If $1 \leq p < \infty$ and $\lambda > n$, then we get $L^{p,\lambda} = \{ 0 \}$.

In this note, we investigate the boundedness of Hardy-Littlewood maximal operator, the singular integral operator of Calderon-Zygmund type and associated maximal singular integral operator on the generalized Morrey spaces $L^{p,\Phi}$. The letters C's will denote positive constants, which may have different values in each line.

2. Hardy-Littlewood maximal operator.

Let f be a locally integrable function on R^n. The Hardy-Littlewood maximal operator, $M : f \to Mf$, is defined by

$$Mf(x) = \sup |B|^{-1} \int_B |f(y)| dy$$

where the supremum is taken over all balls B containing x and $|B|$ is the volume of the ball B. Let

$$M_q f(x) = \{(M|f|^q)(x)\}^{1/q}$$

for $1 \le q < \infty$. We begin by the following.

Theorem 2.1. *Let $f \in L^{p,\Phi}$ and the doubling constant D satisfy $1 \le D < 2^n$. Let $1 \le p < \infty$ and $1 \le q < \log 2^n / \log D$. Then there exists a constant C depending only on p and q such that*

$$\int_{R^n} |f(x)|^p (M_q \chi_{B(x_0,r)})(x) dx \le C\|f\|_{p,\Phi}^p \Phi(r)$$

for all $f \in L^{p,\Phi}$ and balls $B(x_0, r)$, where $\chi_{B(x_0,r)}$ is the characteristic function of $B(x_0, r)$.

Proof. Let $1 \le p < \infty$ and $f \in L^{p,\Phi}$. Then we get, by a standard estimate of maximal functions $M_q \chi_{B(x_0,r)}$,

$$\int_{R^n} |f(x)|^p (M_q \chi_{B(x_0,r)})(x) dx$$

$$= \int_{B(x_0,2r)} |f(x)|^p \{(M|\chi_{B(x_0,r)}|^q)(x)\}^{1/q} dx$$

$$+ \sum_{k=1}^{\infty} \int_{B(x_0,2^{(k+1)}r) - B(x_0,2^k r)} |f(x)|^p \{(M|\chi_{B(x_0,r)}|^q)(x)\}^{1/q} dx$$

$$\le \int_{B(x_0,2r)} |f(x)|^p dx + \sum_{k=1}^{\infty} (2^{1-k})^{n/q} \int_{B(x_0,2^{k+1}r)} |f(x)|^p dx$$

$$\le C\|f\|_{p,\Phi}^p \{\Phi(2r) + \sum_{k=1}^{\infty} (2^{1-k})^{n/q} \Phi(2^{k+1}r)\}$$

$$\le C\|f\|_{p,\Phi}^p \sum_{k=0}^{\infty} (2^{-k})^{n/q} \Phi(2^{k+1}r)$$

$$\le C\|f\|_{p,\Phi}^p \sum_{k=0}^{\infty} (2^{-k})^{n/q} D^{k+1} \Phi(r) = C\|f\|_{p,\Phi}^p \Phi(r),$$

where we used the doubling condition (1) of Φ and the fact that the series $\sum_{k=0}^{\infty} (D2^{-n/q})^k$ is convergent since $1 \le q < \log 2^n / \log D$ and $1 \le D < 2^n$.

$$\text{Q.E.D.}$$

Theorem 2.2. *Let $1 \le D(\Phi) < 2^n$.*
(i) *If $1 < p < \infty$ and $1 \le q < p < \infty$, then M_q is bounded on $L^{p,\Phi}$, that is, there exists a constant C, depending only on p and q, such that*

$$\|M_q f\|_{p,\Phi} \le C\|f\|_{p,\Phi}$$

for all $f \in L^{p,\Phi}$.

(ii) *If $1 \leq p < \infty$, then there exists a constant C, depending only on p, such that*

$$|\{M_p f > t\} \cap B(x_0, r)| \leq Ct^{-p}\|f\|_{p,\Phi}^p \Phi(r)$$

for all $f \in L^{p,\Phi}$.

Proof. We use the method by F.Chiarenza and M.Frasca [3].

(i) In the case $1 < p < \infty$, we recall the weighted maximal inequality established by C.L.Fefferman and E.M.Stein [5]; there exists a constant C such that

$$\int_{R^n} \{(Mf)(x)\}^p \phi(x)dx \leq C \int_{R^n} |f(x)|^p (M\phi)(x)dx \tag{3}$$

for any functions f and $\phi \geq 0$ on R^n. Let $f \in L^{p,\Phi}$ and B a ball. We take $\phi(x)$ as the characteristic function $\chi_B(x)$ of the ball $B = B(x_0, r)$. Then we get by (3),

$$\int_B \{(M_q f)(x)\}^p dx \leq C \int_{R^n} |f(x)|^p (M\chi_B)(x)dx.$$

Then Therorem 2.1 with $1 < p < \infty$ and $q = 1$ implies that the right hand side is majorized by $\|f\|_{p,\Phi}^p \Phi(r)$. Thus

$$\int_{B(x_0,r)} \{(M_q f)(x)\}^p dx \leq C\|f\|_{p,\Phi}^p \Phi(r),$$

that is, $M_q f \in L^{p,\Phi}$ and $\|M_q f\|_{p,\Phi} \leq C\|f\|_{p,\Phi}$.

(ii) In the case $1 \leq p < \infty$, we use the corresponding weak type inequality, that is, there exists a constant C such that for all $f, \phi \geq 0$ and $t > 0$,

$$\int_{\{x:Mf(x)>t\}} \phi(x)dx \leq Ct^{-1} \int_{R^n} |f(x)|(M\phi)(x)dx \tag{4}$$

where C is independent of $f, \phi \geq 0$ and $t > 0$.

Then we get, by (4) and Theorem 2.1 with $1 \leq p < \infty$ and $q = 1$,

$$|\{M_p f > t\} \cap B(x_0, r)| = \int_{\{M(|f|^p)>t^p\}} \chi_{B(x_0,r)}(x)dx$$

$$\leq Ct^{-p} \int_{R^n} |f(x)|^p (M\chi_{B(x_0,r)})(x)dx \leq Ct^{-p}\|f\|_{p,\Phi}^p \Phi(r).$$

Q.E.D.

3. Singular integral operators.

In this section we show the boundedness of Calderon-Zygmund operator on the space $L^{p,\Phi}$.

Definition 3.1. *Let T be the singular integral operator $Tf = k * f$ defined by the kernel k satisfying the conditions,*

$$\|\hat{k}\|_\infty \leq C,$$

$$|k(x)| \leq C|x|^{-n} \quad for \ 0 \neq x \in R^n,$$

$$|k(x) - k(x - y)| \leq C|y|/|x|^{n+1} \quad for \ |y| \leq |x|/2.$$

For $\epsilon > 0$, put

$$T_\epsilon f(x) = \int_{|y| > \epsilon} k(y) f(x - y) dy \quad and \quad T^* f(x) = \sup_{\epsilon > 0} |T_\epsilon f(x)|.$$

Theorem 3.2. Let $1 \leq D(\Phi) < 2^n$.

(i) If $1 < p < \infty$, then the singular integral operator T is bounded on $L^{p, \Phi}$.

(ii) If $p = 1$, then a weak type inequality of T is valid, more precisely, there exists a constant C such that

$$|\{|Tf| > t\} \cap B(x_0, r)| \leq Ct^{-1} \|f\|_{1, \Phi} \Phi(r)$$

where C is independent of $f \in L^{1, \Phi}$ and balls $B(x_0, r)$.

Moreover, (i) and (ii) are valid for the associated maximal operator T^* in place of T.

Proof. (i) In the case $1 < p < \infty$, we use the maximal inequality due to A.Cordoba and C.L.Fefferman [4] (see J.Garcia-Cuerva and J.L.Rubio de Francia [7]) ; there exists a constant C depending only on T, p and $1 < q < \infty$, such that

$$\int_{R^n} |Tf(x)|^p \phi(x) dx \leq C \int_{R^n} |f(x)|^p (M_q \phi)(x) dx \tag{5}$$

for every f and $\phi(x) \geq 0$. (5) is valid for T^* in place of T.

Taking $\phi(x)$ as $\chi_B(x)$ the characteristic function of a ball $B = B(x_0, r)$, we get

$$\int_B |Tf(x)|^p dx \leq C \int_{R^n} |f(x)|^p (M_q \chi_B)(x) dx$$

for some constant C independent of $f \in L^{p, \Phi}$ and $B = B(x_0, r)$.

Choose q so that $1 < q < \log 2^n / \log D$. Then we obtain (i) by Theorem 2.1.

The corresponding inequality for T^* is obvious.

(ii) In the case $p = 1$, it suffices to recall the inequality,

$$\int_{\{T^* f(x) > t\}} \phi(x) dx \leq Ct^{-1} \int_{R^n} |f(x)| (M_q \phi)(x) dx$$

where $1 < q < \infty$ and C is independent of f, $\phi \geq 0$ and $t > 0$.

See J.Garcia-Cuerva and J.L.Rubio de Francia [7; p.504, Corollary 4.11 and p.158, Theorem 3.4.].

Q.E.D.

4. Generalized Campanato-Stampacchia spaces $\tilde{L}^{p,\Phi}$.

J.Peetre [11] investigated the following function spaces $\tilde{L}^{p,\Phi}$ related to the singular integral operator of Calderon-Zygmund type, where $1 \leq p < \infty$ and $\Phi(r)$ is the same growth function introduced by (1) in Section 1.

$\tilde{L}^{p,\Phi} = \tilde{L}^{p,\Phi}(R^n)$ *is the space of locally integrable functions* $f(x)$ *such that*

$$\inf_{\sigma} \int_{B(x_0,r)} |f(x) - \sigma|^p dx \leq C^p \Phi(r) \tag{6}$$

for all balls $B(x_0, r)$, *where the infimum is taken over all complex numbers* σ.

Let $\|f\|_{p,\Phi}^*$ *be the smallest constant* C *satisfying* (6). *Then* $\tilde{L}^{p,\Phi}$ *is a Banach space modulo constants.*

An elementary calculation implies that the semi-norm $\|f\|_{p,\Phi}^*$ is equivalent to

$$\sup_{B} \{\Phi(r)^{-1} \int_B |f(x) - f_B|^p dx\}^{1/p},$$

where f_B is the average of f over the ball $B = B(x_0, r)$. The generalized Morrey space $L^{p,\Phi}$ is continuously embedded in the generalized Campanato-Stampacchia space $\tilde{L}^{p,\Phi}$.

If $\Phi(r) = r^\lambda, 0 \leq \lambda$, we write simply $\tilde{L}^{p,\lambda}$ for $\tilde{L}^{p,r^\lambda}$.

J.Peetre [11] pointed out that, if $0 < \lambda < n$, the Banach space $\tilde{L}^{p,\lambda}$, modulo constants, is equivalent to the classical Morrey space $L^{p,\lambda}$. See also J.Peetre [12].

Here we give a proof.

Proposition 4.1. *Let* $0 < \lambda < n$ *and* $1 \leq p < \infty$. *If* $f \in \tilde{L}^{p,\lambda}$, *then there exists a constant* $\sigma = \sigma(f)$ *such that*

$$\sigma(f) = \lim_{r \to +\infty} f_{B(x_0,r)},$$

which is independent of x_0, *and*

$$\|f - \sigma\|_{p,\lambda} \leq C\|f\|_{p,\lambda}^*$$

for some constant C *independent of* f.

Conversely, if $f \in L^{p,\lambda}$, *then* $f \in \tilde{L}^{p,\lambda}$, *and*

$$\|f\|_{p,\lambda}^* \leq C\|f\|_{p,\lambda}$$

for some constant C *independent of* f.

The second part of the Proposition is obvious. To show the first part we observe the following. The proof depends on a simple Lemma. Confer [2] and [8].

Lemma 4.2. *Suppose* $f \in \tilde{L}^{p,\lambda}, \lambda \geq 0$. *Let* $B_0 = B(x_0, r_0)$ *and* $B_1 = B(x_1, r_1)$ *be balls such that* $B_0 \subset B_1$ *and* $r_1 \leq 2r_0$. *Then*

$$|f_{B_0} - f_{B_1}| \leq C\|f\|_{p,\lambda}^* r_0^{(\lambda-n)/p} \tag{7}$$

for some constant C independent of f and B_0.

Proof of Lemma 4.2. An elementary calculation gives

$$|f_{B_0} - f_{B_1}| \leq \{|B_0|^{-1} \int_{B_0} |f(x) - f_{B_1}|^p dx\}^{1/p}.$$

Since $B_0 \subset B_1$ and $r_0 \sim r_1$, the last term is bounded by $C\|f\|_{p,\lambda}^* r_0^{(\lambda-n)/p}$.

Q.E.D.

Proof of Proposition 4.1. First we prove the existence of $\sigma(f)$. Let $B_0 = B(x_0, 2^k)$ and $B_1 = B(x_0, 2^{k+1})$, then by (7)

$$|f_{B_0} - f_{B_1}| \leq C\|f\|_{p,\lambda}^* 2^{k(\lambda-n)/p}$$

for all $x_0 \in R^n, k$ and f. Similarly, if $2^k < r \leq 2^{k+1}$, we have

$$|f_{B(x_0,r)} - f_{B(x_0,2^k)}| \leq C\|f\|_{p,\lambda}^* r^{(\lambda-n)/p}$$

Consequently, there exists a limit, $\sigma(f, x_0) = \lim_{r \to +\infty} f_{B(x_0,r)}$, and

$$|\sigma(f, x_0) - f_{B(x_0,r)}| \leq C\|f\|_{p,\lambda}^* r^{(\lambda-n)/p}.$$

If $x_0 \neq x_1$, then let $B_0 = B(x_0, r)$ and $B_1 = B(x_1, 2r)$ with $r > |x_0 - x_1|$. Then $B_0 \subset B_1$ and from Lemma 4.2 again, we get

$$|f_{B_0} - f_{B_1}| \leq C\|f\|_{p,\lambda}^* r^{(\lambda-n)/p},$$

which tends to 0 as $r \to \infty$. Thus $\sigma(f, x_0) = \sigma(f, x_1)$. Thus the limit $\sigma(f, x_0)$ is independent of x_0 and

$$|\sigma(f) - f_{B(x_0,r)}| \leq C\|f\|_{p,\lambda}^* r^{(\lambda-n)/p}. \tag{8}$$

We have

$$\{r^{-\lambda} \int_{B(x_0,r)} |f(x) - \sigma|^p dx\}^{1/p}$$

$$\leq \{r^{-\lambda} \int_{B(x_0,r)} |f(x) - f_{B(x_0,r)}|^p dx\}^{1/p} + \{r^{-\lambda} \int_{B(x_0,r)} |f_{B(x_0,r)} - \sigma|^p dx\}^{1/p}.$$

The first term on the right hand side is bounded by $C\|f\|_{p,\lambda}^*$ and the second has the same bound by (8).

Q.E.D.

References.

[1] D.R.Adams, A note on Riesz potentials, Duke Math. J. 42(1975), 765-778.

[2] S.Campanato, Proprieta di una famiglia di spazi funzionali, Ann. Scuola Norm. Sup. Pisa, 18(1964), 137-160.

[3] F.Chiarenza and M.Frasca, Morrey spaces and Hardy-Littlewood maximal function, Rend. Mat.(7), 7(1987), 273-279.

[4] A.Cordoba and C.L.Fefferman, A weighted norm inequality for singular integrals, Studia Math. 57(1976), 97-101.

[5] C.L.Fefferman and E.M.Stein, Some maximal inequalities, Amer. J. Math. 93(1971), 107-115.

[6] C.L.Fefferman and E.M.Stein, H^p spaces of several variables, Acta Math. 129(1972), 127-193.

[7] J.Garcia-Cuerva and J.L.Rubio de Francia, Weighted norm inequalities and related topics, North-Holland, 1985.

[8] S.Janson, M.H.Taiblesom and G.Weiss, Elementary characterization of the Morrey-Campanato spaces, Proc. in Harmonic Analysis, Cortona, Italy, Springer, Lecture Notes in Math. Vol.992, 1982, 101-114.

[9] R.L.Long, The spaces generated by blocks, Sci. Sinica, Ser.A, 27(1984), 16-26.

[10] C.B.Morrey,Jr. On the solutions of quasi-linear elleptic partial differential equations, Trans. A.M.S. 43(1938), 126-166.

[11] J.Peetre, On convolution operators leaving $L^{p,\lambda}$ spaces invariant, Ann. Mat. Pura Appl.(4), 72(1966), 295-304.

[12] J.Peetre, On the theory of $L_{p,\lambda}$ spaces, J. Funct. Anal. 4(1969), 71-87.

INTERPOLATION OF SPACES DEFINED BY THE LEVEL FUNCTION†

Gord Sinnamon‡
University of Western Ontario
London, Ontario, CANADA

Section 1: Introduction.

Suppose λ is a regular, Borel measure on $\mathbf{R}$ and suppose that $\lambda(-\infty, x) < \infty$ for all $x \in \mathbf{R}$. The Lebesgue spaces, L^p_λ, for $1 \le p \le \infty$ will then contain non-trivial, non-increasing functions. Define

$$\|f\|_{p\downarrow\lambda} = \sup\left\{\int_{-\infty}^{\infty} |f|g\,d\lambda : g \ge 0, g \text{ non-increasing}, \|g\|_{p',\lambda} \le 1\right\}$$

where p' is defined by $1/p + 1/p' = 1$.

The norm $\|\cdot\|_{p\downarrow\lambda}$ was utilized in the early 1950's by Halperin [2] and Lorentz [3] and more recently, in proving weighted norm inequalities, by Sinnamon [5], Neugebauer [4], Stepanov [7] and others. Halperin showed, in the case of λ absolutely continuous, that to each non-negative function f there corresponds a function f^o which is non-increasing and satisfies $\|f\|_{p\downarrow\lambda} = \|f^o\|_{p,\lambda}$. He called (a variant of) this function f^o, the level function of f. In [6] this contruction is extended to regular, Borel measures and the dual of the Banach space $L^{p\downarrow}_\lambda$ is characterised for $1 \le p < \infty$. The dual space is the space $L^{p'}_\lambda{}^*$, with norm $\|f\|_{p'*\lambda} = \|\bar{f}\|_{p',\lambda}$ where

$$\bar{f}(x) = \operatorname{ess\,sup}_{y \ge x} |f(y)|.$$

In this paper the above duality result is used to show that interpolation between the spaces $L^{p\downarrow}_\lambda$, $1 \le p < \infty$, again yields the spaces $L^{p\downarrow}_\lambda$.

We conclude this section with some definitions and notation. The main results of the paper are contained in Section 2.

Let f be a λ-measurable function. The distribution function of f with respect to the measure λ is $m_f(\alpha) = \lambda\{x : |f(x)| > \alpha\}$. The non-increasing rearrangement of f is $f^*(t) = \inf\{\alpha : m_f(\alpha) \le t\}$, $t > 0$. The following simple facts follow directly

---

† This paper is in final form and no version of it has been or will be submitted elsewhere. Revised June 5, 1991.

‡ Research supported by a grant from the Natural Sciences and Engineering Research Council of Canada.

from the definitions. 1. $m_f(f^*(t)) \le t$. 2. $f^*(m_f(\alpha)) \le \alpha$. 3. $f^*(0) = \|f\|_{\infty,\lambda}$. 4. If $|f| \le |f_0| + |f_1|$ and $0 < \varepsilon < 1$ then $f^*(s) \le f_0^*((1-\varepsilon)s) + f_1^*(\varepsilon s)$.

If $1 \le p < \infty$, L_λ^p is the space of all λ-measurable functions for which the norm $\|f\|_{p,\lambda} = (\int_{-\infty}^{\infty} |f|^p \, d\lambda)^{1/p}$ is finite. L_λ^∞ is the space of all essentially bounded functions. The norm on the space is $\|f\|_{\infty,\lambda} = \operatorname{ess\,sup}|f|$.

Reference is made to several specific theorems in [1]. That text will also serve as our basic reference in interpolation theory and in particular we will follow the notation used there.

Section 2: Main Results.

The interpolation results are first proved in the dual spaces $L_\lambda^{p'}{}^*$. To do this, a formula for the K-functional, $K(t, f; L_\lambda^1{}^*, L_\lambda^\infty{}^*)$, is derived and then the reiteration theorem and the results of ordinary L_λ^p interpolation are applied. The Duality Theorem for Real Interpolation provides the link to the spaces $L_\lambda^{p\downarrow}$.

We begin with a simple fact.

Lemma 1. $L_\lambda^\infty{}^* = L_\lambda^\infty$ with identical norms.

Proof. Suppose f is λ-measurable. Since $f \le \bar{f}$ almost everywhere we certainly have $\|f\|_{\infty,\lambda} \le \|\bar{f}\|_{\infty,\lambda} = \|f\|_{\infty * \lambda}$. Conversely, for each x, $\bar{f}(x) = \operatorname{ess\,sup}_{y \ge x} |f(y)| \le \|f\|_{\infty,\lambda}$ so $\|f\|_{\infty * \lambda} = \|\bar{f}\|_{\infty,\lambda} \le \|f\|_{\infty,\lambda}$.

The following theorem gives a formula for the K-functional for the pair $(L_\lambda^1{}^*, L_\lambda^\infty{}^*)$.

Theorem 1. $K(t, f; L_\lambda^1{}^*, L_\lambda^\infty{}^*) = \int_0^t \bar{f}^*(s) \, ds = K(t, \bar{f}; L_\lambda^1, L_\lambda^\infty)$.

Proof. Fix a λ-measurable function f and fix $t > 0$. The second equality above is from [1, Theorem 5.2.1]. To establish the first we prove inequalities in both directions beginning with $K(t, f, L_\lambda^1{}^*, L_\lambda^\infty{}^*) \ge \int_0^t \bar{f}^*(s) \, ds$.

Suppose $f = f_0 + f_1$. Clearly $\bar{f} \le \bar{f}_0 + \bar{f}_1$ so for each $\varepsilon > 0$ we have $\bar{f}^*(s) \le \bar{f}_0^*((1-\varepsilon)s) + \bar{f}_1^*(\varepsilon s)$ and hence

$$\int_0^t \bar{f}^*(s) \, ds \le \int_0^t \bar{f}_0^*((1-\varepsilon)s) \, ds + \int_0^t \bar{f}_1^*(\varepsilon s) \, ds.$$

The second integral on the right hand side is dominated by

$$t \bar{f}_1^*(0) = t \|\bar{f}_1\|_{\infty,\lambda} = t \|f_1\|_{\infty * \lambda}.$$

The first is dominated by

$$\int_0^\infty \bar{f}_0^*((1-\varepsilon)s) \, ds = \frac{1}{1+\varepsilon} \int_0^\infty \bar{f}_0^*(s) \, ds = \frac{1}{1+\varepsilon} \int_{-\infty}^\infty \bar{f}_0(x) \, dx = \frac{1}{1+\varepsilon} \|f_0\|_{1 * \lambda}.$$

Letting $\varepsilon \to 0$ we have

$$\int_0^t \bar{f}^{\,*}(s)\,ds \le \|f_0\|_{1\,*\,\lambda} + t\|f_1\|_{\infty\,*\,\lambda}.$$

Since $K(t, f; L_\lambda^{1*}, L_\lambda^{\infty*}) = \inf_{f_0+f_1=f}(\|f_0\|_{1\,*\,\lambda} + t\|f_1\|_{\infty\,*\,\lambda})$ we have the inequality.

To prove the other inequality, that $K(t, f, L_\lambda^{1*}, L_\lambda^{\infty*}) \le \int_0^t \bar{f}^{\,*}(s)\,ds$, we set $a = \bar{f}^{\,*}(t)$ and $E = \{x : |\bar{f}(x)| > a\}$. Note that $\lambda(E) = m_{\bar{f}}(a) = m_{\bar{f}}(\bar{f}^{\,*}(t)) \le t$. Also, if $\lambda(E) \le s \le t$ then $\bar{f}^{\,*}(t) \le \bar{f}^{\,*}(s) \le \bar{f}^{\,*}(\lambda(E)) = \bar{f}^{\,*}(m_{\bar{f}}(a)) \le a = \bar{f}^{\,*}(t)$ so $\bar{f}^{\,*}(s) = \bar{f}^{\,*}(t)$. Thus $\bar{f}^{\,*}$ is constant on the interval $[\lambda(E), t]$.

Set $f_0(x) = \max(0, |f(x)| - a)\mathrm{sgn}(f(x))$. Here $\mathrm{sgn}(z) = z/|z|$ if $z \ne 0$ and $\mathrm{sgn}(z) = 0$ otherwise. $\bar{f}_0$ is related to $\bar{f}$ in a simple way.

$$\bar{f}_0(x) = \mathrm{ess\,sup}_{y \ge x}|f_0(y)| = \mathrm{ess\,sup}_{y \ge x}\max(0, |f(y)| - a)$$

$$= \max(0, \mathrm{ess\,sup}_{y \ge x}|f(y)| - a) = \max(0, \bar{f}(x) - a).$$

The remaining portion of f is bounded by a. Let $f_1 = f - f_0$ and we have

$$|f_1(x)| = |f(x) - f_0(x)| = |f(x)| - \max(0, |f(x)| - a) \le a.$$

Thus $\|f_1\|_{\infty,\lambda} \le a$. We can now estimate the K-functional.

$$K(t, f; L_\lambda^{1*}, L_\lambda^{\infty*}) \le \|f_0\|_{1\,*\,\lambda} + t\|f_1\|_{\infty\,*\,\lambda}.$$

The set E, defined earlier, contains the support of $\bar{f}_0$. $\bar{f}_0$ takes the value $\bar{f}(x) - a$ on E. This fact, together with Lemma 1 shows that the K-functional is dominated by

$$\int_{-\infty}^{\infty} \bar{f}_0(x)\,dx + t\|f_1\|_{\infty,\lambda} \le \int_E \bar{f}(x) - a\,dx + ta \le \int_0^{\lambda(E)} \bar{f}^{\,*}(s) - a\,ds + ta.$$

We now recall that $\bar{f}^{\,*}$ takes the value a on the interval $[\lambda(E), t]$ so the last expression is just

$$\int_0^t \bar{f}^{\,*}(s) - a\,ds + ta = \int_0^t \bar{f}^{\,*}(s)\,ds.$$

This completes the proof.

With the above relation between the K-functionals for $(L_\lambda^{1*}, L_\lambda^{\infty*})$ and $(L_\lambda^1, L_\lambda^\infty)$ we can prove the following.

Corollary 1. *If $1 \le p_0 < p_1 \le \infty$, $0 < \theta < 1$, and $1/p = (1-\theta)/p_0 + \theta/p_1$ then $(L_\lambda^{p_0*}, L_\lambda^{p_1*})_{\theta,p} = L_\lambda^{p*}$ with equivalent norms.*

Proof.

$$\|f\|_{(L_\lambda^{1*},L_\lambda^{\infty*})_{1/p',p}} = \left(\int_0^\infty \left(t^{-1/p'}K(t,f;L_\lambda^{1*},L_\lambda^{\infty*})\right)^p t^{-1}\,dt\right)^{1/p}$$

$$= \left(\int_0^\infty \left(t^{-1/p'}K(t,\bar{f};L_\lambda^1,L_\lambda^\infty)\right)^p t^{-1}\,dt\right)^{1/p} = \|\bar{f}\|_{(L_\lambda^1,L_\lambda^\infty)_{1/p',p}}.$$

By [1, Theorem 5.2.1] we have $(L_\lambda^1,L_\lambda^\infty)_{1/p',p} = L_\lambda^p$ with equivalent norms. Therefore the norm on $(L_\lambda^{1*},L_\lambda^{\infty*})_{1/p',p}$ is equivalent to $\|\bar{f}\|_{p,\lambda} = \|f\|_{p*\lambda}$. Using this fact and the reiteration theorem ([1, Theorem 3.5.3]) we obtain

$$(L_\lambda^{p_0*},L_\lambda^{p_1*})_{\theta,p} = ((L_\lambda^{1*},L_\lambda^{\infty*})_{\frac{1}{(p_0)'},p_0},(L_\lambda^{1*},L_\lambda^{\infty*})_{\frac{1}{(p_1)'},p_1})_{\theta,p} = (L_\lambda^{1*},L_\lambda^{\infty*})_{1/p',p} = L_\lambda^{p*}.$$

with equivalent norms. Here $(L_\lambda^{1*},L_\lambda^{\infty*})_{0,1}$ is taken to be L_λ^{1*} and $(L_\lambda^{1*},L_\lambda^{\infty*})_{1,\infty}$ is taken to be $L_\lambda^{\infty*}$.

Corollary 2. *If* $1 \le p_0 < p_1 < \infty$, $0 < \theta < 1$, *and* $1/p = (1-\theta)/p_0 + \theta/p_1$ *then* $(L_\lambda^{p_0\downarrow},L_\lambda^{p_1\downarrow})_{\theta,p} = L_\lambda^{p\downarrow}$ *with equivalent norms.*

Proof. As mentioned, the dual space of $L_\lambda^{p\downarrow}$ is $L_\lambda^{p'*}$ for $1 \le p < \infty$. Thus, by Corollary 1 and the Duality Theorem for Real Interpolation [1, Theorem 3.7.1],

$$(L_\lambda^{p\downarrow})' \equiv L_\lambda^{p'*} = (L_\lambda^{p'_0*},L_\lambda^{p'_1*})_{\theta,p'} \equiv ((L_\lambda^{p_0\downarrow})',(L_\lambda^{p_1\downarrow})')_{\theta,p'} = (L_\lambda^{p_0\downarrow},L_\lambda^{p_1\downarrow})'_{\theta,p}$$

with equivalent norms. Here "$\equiv$" indicates the isomorphism $f \leftrightarrow L_f$ where $L_f(g) = \int_{\mathbf{R}} fg\,d\lambda$. Since all the $L_\lambda^{p\downarrow}$ spaces for $1 \le p < \infty$ have a common dense subset, ($\{f : f$ is bounded and supported on $(-\infty,x]$ for some $x\}$ will do,) it follows that $L_\lambda^{p\downarrow} = (L_\lambda^{p_0\downarrow},L_\lambda^{p_1\downarrow})_{\theta,p}$. This completes the proof.

References

1. J. Bergh and J. Löfström. *Interpolation Spaces, An Introduction.* Berlin, Heidelberg 1976, Springer-Verlag.
2. I. Halperin. Function spaces. *Canad. J. Math.* **5** (1953), 278-288.
3. G. G. Lorentz. *Bernstein Polynomials.* Toronto 1953, University of Toronto Press.
4. C. J. Neugebauer. Weighted norm inequalities for averaging operators of monotone functions. Preprint.
5. G. Sinnamon. Weighted Hardy and Opial-type inequalities. *J. Math. Anal. Appl.* To appear.
6. G. Sinnamon. Spaces defined by the level function and their duals. Preprint.
7. V. Stepanov. The weighted Hardy's inequality for nonincreasing functions. Preprint.

GROUPS OF SUPERPOLYNOMIAL GROWTH ·

N. Th. Varopoulos

Université Paris VI - Analyse Complexe et Géométrie (URA 213)

4, Place Jussieu, 75252 Paris cedex 05

1. Distance and volume growth.

Let G be a discrete group generated by a finite number of generators $\gamma_1, \ldots, \gamma_k \in G$. One defines then a distance $d(\cdot, \cdot)$ on G by requiring that $d(gx, gy) = d(x, y)$ $(x, y, g \in G)$ and that $d(e, x)$, the distance of $x \in G$ from the neutral point $e \in G$ is, by definition, the smallest $n \geq 0$ for which we can write $x = \gamma_{i_1}^{\epsilon_1} \cdots \gamma_{i_n}^{\epsilon_n}$, $(i_1, \ldots, i_n = 1, \ldots, k\,;\, \epsilon_j = 0, \pm 1)$.

Let G be a connected real Lie group and let $X_1, \ldots, X_k \in \mathcal{L}(G)$ be a finite number of generators of the Lie algebra of G ; in other words $X_1, \ldots, X_k$ are left invariant C^∞ fields on G that together with their successive brackets $[X_{i_1}[X_{i_2}, \ldots, X_{i_s}] \ldots]$ generate the tangent space. We say that an absolutely continuous path $\ell(t) \in G$ $(0 \leq t \leq T)$ is of length less or equal to T if its speed vector $\dot{\ell}(t) = d\ell(\dfrac{\partial}{\partial t})$ (with respect to $X_1, \ldots, X_k$) is almost everywhere length ≤ 1 : this means that $\dot{\ell}(t) = \sum_{j=1}^{k} a_j X_j$ (p.p.t ; $\sum a_j^2 \leq 1$). We then say that $d(x, y) \leq T$ $(x, y \in G)$ if we can join x to y with a path of length $\leq T$.

The growth function $\gamma(t)$ $(t > 0)$ of G, when G is either a discrete or a Lie group, is defined to be : $\gamma(t) = $ the Haar measure of a ball of radius t. For large t $(t \geq 1)$ the above function $\gamma(t)$ is essentially independent of the particular choice of the generators used : $\gamma(t)$ $(t \geq 1)$ is thus a group invariant. For Lie groups and $0 < t < 1$ the behaviour of $\gamma(t)$ does depend on the choice of $X_1, \ldots, X_k$ but we always have $\gamma(t) \approx t^\delta$ with $\delta = \delta(G, X_1, X_2, \ldots, X_k) = 1, 2, \ldots$ (this is a theorem of Nagel-Stein-Wainger). For $t \geq 1$ and a Lie group we have either $\gamma(t) \approx t^D$ with $D = D(G) = 0, 1, \ldots$ or $\gamma(t) \geq C e^{ct}$ (this is a theorem of Guivarc'h). In the discrete case we have either $\gamma(t) \approx t^D$ $(D(G) = 0, 1, 2, \ldots)$, if G is a finite extension of a nilpotent group or $\gamma(t) t^{-A} \to \infty$ for all $A \geq 1$ in all other cases (this is a theorem of Gromov).

2. The diffusion and the random walks.

Let G be a unimodular Lie group with $X_1, \ldots, X_k \in \mathcal{L}(G)$ as above we can then consider $\Delta = -\sum X_j^2$ which can be identified with a self adjoint (positive) operator on $L^2(G)$ and we can also consider $T_t = \exp(-t\Delta)$ the corresponding submarkovian semigroup. The kernel of that semigroup will be denoted by $p_t(x, y) = p_t(x^{-1}y)$ $(t > 0; x, y \in G)$. The discrete analogue of the above diffusion is of course the random walk defined on a discrete group by the transition matrix $M(x, y) = \mu(x^{-1}y)$ $(x, y \in G)$ where $\mu \in P(G)$ is a symmetric probability measure on G. We shall consider in what follows, essentially, only random walks that are defined by symmetric measures that have *generating* supports ($: Gp(\text{supp } \mu) = G$). What we shall examine then is the convolution power μ^n of that measure or equivalently $T_t = \exp(-t(\delta - \mu))$ the continuous time Markov semigroup that it generates.

3. The continuous time semigroups.

Let $M(t) \in \mathbf{R}$ $(t > 0)$ be a continuous non positive decreasing function of t s.t. $M(t) = O(t)$ $(t \to \infty)$, $|M(2t)/M(t)| \leq c$ for some fixed $c > 0$. Let also $\mu \in P(G)$ be a symmetric probability measure on the discrete group as above.

Proposition 1.

Let us assume that $\mu^n(e) = O(e^{M(t)})$. Then $T_t(e) = O(e^{cM(ct)})$. Conversely let us assume that $T_t(e) = O(e^{M(t)})$. Then $\mu^n(e) = O(e^{cM(ct)})$.

All the C's and c's that appear here and in what follows are positive constants independent of all the important parameters but which vary from place to place.

The proof is an easy consequence of the fact that $T_t(e)$ is decreasing in t, that $\mu^n(e)$ can be assumed decreasing in n (by replacing $\mu^n(e)$ by $\mu^{2n}(e) = \|\mu^n\|_2^2$) and finally of the elementary estimates on the coefficients $c_n = e^{-t}\dfrac{t^n}{n!}$ of $T_t = \sum c_n \mu^n$ that imply that

$$\sum_{n=1}^{[ct]} c_n = O(e^{-t/2}) \text{ if } c > 0 \text{ is small enough and } \sum_{n < [Ct]} c_n > 1/2 \text{ if } C > 0 \text{ is large enough.}$$

We have for instance :

$$T_t(e) \leq \sum_{n=0}^{[ct]} c_n + e^{M(ct)} \sum_{n > [ct]} c_n \leq O(e^{-t/2}) + O(e^{M(ct)}).$$

Conversely we have

$$\mu^{2[Ct]}(e) \sum_{n \leq Ct} c_{2n} \leq T_t(e).$$

Let us now quite generally assume that $T_t = e^{-tA}$ is a symmetric submarkovian semigroup on $L^2(X)$ ((X, dx) is here an abstract measure space). We shall consider then :

$$P_t = e^{-tA^{1/2}} = \frac{1}{\sqrt{\pi}} \int_0^\infty \frac{e^{-u}}{\sqrt{u}} T_{t^2/4u} \, du$$

the subordinated semigroup (cf. [1]). Let us fix $0 < \alpha \leq 1$. For any $x \in X$ we have then :

(1) $$p_t(x, x) = O(\exp(-ct^\alpha)) \Leftrightarrow q_t(x, x) = O(\exp(-ct^{\frac{2\alpha}{\alpha+1}}))$$

where p_t (resp. q_t) denotes the kernel of T_t (resp. P_t).

Proof.

We have :

$$q_t = \frac{1}{\sqrt{\pi}} \int_0^\lambda + \int_\lambda^\infty \frac{e^{-u}}{\sqrt{u}} p_{t^2/4u} \, du = I_1 + I_2$$

where

$$I_1 \leq C \exp\left(-c\frac{t^{2\alpha}}{\lambda^\alpha}\right) \; ; \; I_2 \leq Ce^{-\lambda/2} \int_\lambda^\infty \frac{e^{-u/2}}{\sqrt{u}} \, du.$$

If we chose $\lambda = t^{2\alpha/1+\alpha}$ we obtain the implication $\Rightarrow$.

Conversely we have :

$$(k\lambda)^{1/2} e^{-2k\lambda} p_{t^2/4k\lambda}(x, x) \leq C \int_{k\lambda}^{2k\lambda} \frac{e^{-u}}{\sqrt{u}} p_{t^2/4u}(x, x) \leq Cq_t(x, x)$$

because all the functions involved are monotone.

If we set $\lambda = t^{\frac{2\alpha}{1+\alpha}}$ and k sufficiently small but fixed we obtain the implication $\Leftarrow$.

Let finally $T_t = e^{-tA}$ and $R_t = e^{-tB}$ be two symmetric submarkovian semigroups on $L^2(X)$ as above, and let us assume that :

(2) $$\|T_t f\|_\infty \leq e^{M(t)} \|f\|_2 \quad t > 0 \quad f \in L^2(X)$$

where $M(t)$ ($t > 0$) is as in proposition 1. Let us further assume that the corresponding Dirichlet forms satisfy $(Af, f) \leq C(Bf, f)$ ($\forall f \in \text{Dom}(B)$). We can conclude then that :

(3) $$\|R_t f\|_\infty \leq e^{C\overline{M}(ct)} \|f\|_2 \quad t > 0 \quad f \in L^2(X)$$

where

$$\overline{M}(t) = \frac{1}{t} \int_0^t M(t) \, dt.$$

This is a deep result and it followes from the work of E.B. Davies (cf. theorem 2.2.3 and corollary 2.2.8 in [2]).

4. The growth and the potential function.

Let now G be a discrete group as above and let us fix some set of generators and denote by $|x| = d(e, x)$. Let us further assume that $\gamma(n) \geq \exp(cn^{\alpha})$ for some $0 < \alpha \leq 1$.

All non virtually nilpotent soluble groups have exponential growth (i.e. $\gamma(n) \geq \exp(cn)$) : this is a theorem of Milnor). More subtle examples of such groups have recently been given by R. Grigorchuk (cf. [3]) with a growth function that satisfies :

$$\exp(cn^{\beta}) \leq \gamma(n) \leq \exp(cn^{\alpha})$$

with $0 < \beta \leq \alpha < 1$.

We shall now define $\Sigma_n = \{x \in G \,;\, |x| = n\}$ the n-sphere on G and $\chi_n = \chi_{\Sigma_n}$ the characteristic function of Σ_n in G. We shall also define :

(4)
$$\mu = a \sum_{j \geq 0} j^{-2} \sigma(j)^{-1} \chi_j$$

where :

$$\sigma(j) = |\Sigma_j| = \gamma(j) - \gamma(j-1)$$

and where $a > 0$ is chosen so that the above function can be identified with a probability measure i.e. $a \sum j^{-2} = 1$.

The first step towards analysing the above function is to prove that the growth condition :

$$\gamma(n) > \exp(cn^{\alpha})$$

implies that (with the *same* α) :

(5)
$$\sigma(n) \geq |\partial B_n| \geq \exp(cn^{\alpha})$$

where $B_n = \{x \in G \,;\, |x| \leq n\}$ and where ∂A, the boundary of the subset $A \subset G$, is defined by the relation $\partial A = \{x \in A \,;\, d(x, G \setminus A) \leq 1\}$. This non trivial fact follows from the isoperimetric inequalities developed in [4] (cf. also [5] and [6] and the forthcomming book [7]).

Using next the relevant argument in [4] and [6], we obtain that for all $n, p \geq 2$ the convolution powers μ^n of μ satisfy :

$$\mu^n(\{e\}) \leq (1 - s_p)^n + ns_p \exp(-cp^{\alpha})$$

where $s_p = a \sum_{j \geq p} j^{-2} \sim 1/p$.

It follows that :

$$\mu^n(\{e\}) \leq C \exp(-cN) + CN \exp(-cp^\alpha)$$

with $n \gg p$ $N = \dfrac{n}{p}$.

If we chose $p \sim n^{1/1+\alpha}$ we conclude that :

(6) $$\mu^n(\{e\}) = O(\exp(-cn^{\frac{\alpha}{1+\alpha}})).$$

Let now $\nu = \nu^* \in \mathbf{P}(G)$ be a symmetric probability measure on G with generating support. We shall define then $\lambda = (\delta - \nu)^{1/2} = \delta - \nu_1$ which is the generator of the Poisson semigroup $P_t = \exp(-(\delta - \nu)^{1/2}t)$. We have then :

Proposition 2 (Saloff-Coste [8]).

There exists C (that only depends on ν) s.t.

$$2((\delta - \mu)f, f) = \sum_{x,y \in G} \mu(xy^{-1})|f(x) - f(y)|^2 \leq C((\delta - \nu)^{1/2}f, f) \; ; \; \forall f \in c_0(G).$$

This is just the proposition 3 of [8] where we set $\alpha = 1/2$ $\Phi(x) = |x|\mu(x)$.

5. The principal results.

Theorem 1.

Let G be a discrete finitely generated group that satisfies the growth condition $\gamma(n) \geq \exp(cn^\alpha)$ for some fixed $0 < \alpha \leq 1$. Then if $\mu = \mu^* \in \mathbf{P}(G)$ is a probability measure on G with generating support (not necessarily finite support) we have :

$$\mu^n(\{e\}) = O[\exp(-cn^{\frac{\alpha}{\alpha+2}})].$$

The above result is in some sense optimal. In [9] I already gave an example that proves the optimal nature of (7) for a very simple exponential soluble group. More recently in [10] Alexopoulos shows that if G is an exponential polycyclic group, then *every* symmetric finitely supported $\mu \in \mathbf{P}(G)$ satisfies :

$$\mu^{2n}(\{e\}) \geq \exp(-cn^{1/3})$$

and of course we have then $\alpha = 1$.

Let now G be unimodular Lie group and let $X_1, \ldots, X_k$, $\delta \geq 1$, $p_t(x, y)$ and $d(x, y)$ be as in §1, §2. We have :

Theorem 2.

Let G be a unimodular Lie group of exponential growth. For every $\varepsilon > 0$ there exists then $C > 0$ s.t. :

$$p_t(x, y) \leq Ct^{-\delta/2} \exp\left(-ct^{1/3} - \frac{d^2(x, y)}{(4 + \varepsilon)t}\right) \; ; \; x, y \in G, \; t > 0.$$

The analogous "Gaussian" estimate also holds for $\mu^n(x)$.

The above theorems have of course a geometric counter part (in the spirit of [4], [7], [11]) in terms of Orlicz type Sobolev inequalities (cf. [12]). Let me only mention :

Theorem 3.

Let G be a discrete finitely generated group as in theorem 1. Then there exists C s.t. every finite subset with card $A = |A| \geq 2$ satisfies :

$$|\partial A| \geq C|A|(\log|A|)^{-\frac{2}{\alpha}}.$$

Having done what we did in §3, §5 the proof of theorem 1 is easy to give : the measure μ in (4) satisfies (6). The proposition 2 and the Davies result [: (2) $\Rightarrow$ (3)] implies therefore that for every $\nu = \nu^*$ with generating support we have :

$$\exp(-t(\delta - \nu)^{1/2})(\{e\}) = O(\exp[-cn^{\frac{\alpha}{\alpha+1}}]).$$

But (1) and proposition 1 imply then our theorem.

Theorem 3 follows directly by setting $f = \chi_A$ and $\varepsilon \approx$ the appropriate power of $\log|A|$ in (2.2.13) of [2].

In connection with theorem 3 observe that an isoperimetric inequality of the form

(7) $$|\partial A| \geq C|A|(\log|A|)^{-\lambda} \quad A \subset G$$

for some $\lambda \geq 0$ is equivalent to a Sobolev inequality of the form

(8) $$\|\nabla f\|_1 \geq C\varepsilon^\lambda \|f\|_{1+\varepsilon} \; ; \; f \in c_0(G), \; \varepsilon > 0.$$

Indeed this is clear if $f = \chi_A$ is a characteristic function. And (8), for a general f, follows by a standard argument. (8) in turn, by an easy argument, implies a Dirichlet inequality of the form :

(9) $$\|\nabla f\|_2 \geq C\varepsilon^\lambda \|f\|_{2+\varepsilon} \; ; \; f \in c_0(G), \; \varepsilon > 0.$$

(In both in (8) & (9) the C is independent of $\varepsilon > 0$).

It is now a fact (fairly deep) that the estimate (9) implies that the semigroup T_t generated by the Dirichlet form $Q(f,f) = \|\nabla f\|_2^2$ satisfies the estimate

$$\|T_t\|_{1 \to \infty} = O[\exp(-ct^{\frac{1}{2\lambda+1}})].$$

It is also a fact (also fairly deep) that a decay of the form $\mu^n(e) = O(\exp[-cn^\beta])$ implies the estimate (9) with $\lambda = \dfrac{1}{2\beta}$ (cf. [12]). This indicates that the theorem 3 is *not optimal*. Analysing more closely the above circle of ideas we are led to *conjecture* that in theorem 3 we should have (7) with $\lambda = 1/\alpha$. But this is an open problem.

The proof of theorem 2 follows the same lines in the spirit of [4], [7]. The details will be left to the reader. I shall simply observe that in the context of Lie groups what replaces (5) is the estimate :

$$\gamma(t+1) - \gamma(t) \geq \exp(ct^\alpha) \; ; \; t \geq 1.$$

This is a consequence of the Sobolev inequalities of [4], [7] applied to the function $f = \min[1, \operatorname{dist}(G \setminus B_{n+1})]$.

BIBLIOGRAPHY

[1]   K. YOSIDA. Functional analysis, Springer-Verlag.

[2]   E.B. DAVIES. Heat kernel and spectral theory, Cambridge University Press.

[3]   R.I. GRIGORCHUK. Degrees of growth of finitely generated groups and the theory of invariant means, Math. U.S.S.R. Izvestiya, vol.25, 1985, n°2.

[4]   N.Th. VAROPOULOS. J. Funct. Analysis, 76, 1988, p.346-410.

[5]   N.Th. VAROPOULOS. J. Funct. Analysis, 63, 1985, p.215-239.

[6]   N.Th. VAROPOULOS. C.R. Acad. Sci, t.302 (I), 1986, p.203-205.

[7]   N.Th. VAROPOULOS, L. SALOFF-COSTE, T. COULHON. Analysis and Geometry on groups, Cambridge University Press (to appear), Preliminary mimeographied form, Notes Univ. Paris VI.

[8]   L. SALOFF-COSTE. Bull. Sc. Math. 2e série, 113, 1989, p.3-21.

[9]   N.Th. VAROPOULOS. Bull. Sc. Math. 2e série, 107, 1983, p.337-344.

[10]   G. ALEXOPOULOS. C. R. Acad. Sci., t.305 (I), 1987, p.777-779.

[11]   N.Th. VAROPOULOS. Proceedings I.C.M., Kyoto, 1990 (to appear).

[12]   N.Th. VAROPOULOS. J. Funct. Analysis, 63, 1985, p.240-260.

Littlewood-Paley Theory in One and Two Parameters

J. M. Wilson*
Department of Mathematics
University of Vermont
Burlington, Vermont 05405 USA

1. Introduction.

In this talk I shall describe some recent results in Littlewood-Paley theory and their applications to the spectral analysis of certain partial differential operators. I will also sketch a couple of the proofs. As experts in this field know, there are two kinds of Littlewood-Paley theory: "discrete" and "continuous." For the sake of the *non*-experts, I will only discuss the discrete theory, even though all of the applications I will mention are from the continuous theory. All of the important ideas are in the discrete case; readers who are interested in the technical details of going over into the continuous setting will all they want (and perhaps more) in [W2] and [W3].

For the same reason, I will restrict my discussion to dimensions one and two, even though the applications will all be in dimensions three and higher.

Before continuing, I would like thank Professor Igari and the Mathematical Institute of Tohoku University for their kind invitation to speak at this conference. I also wish to thank Tohoku University, the American Mathematical Society, and the National Science Foundation for financial support which made it possible for me to travel to Japan and to Sendai.

2. One parameter.

Let $\mathcal{I}$ denote the family of dyadic intervals on $\mathbf{R}$. To each $I \in \mathcal{I}$ we associate the orthonormalized Haar function h_I with support I. For any locally integrable f we define $\lambda_I(f) = \int f h_I \, dx$. If $f \in L^2(\mathbf{R})$ then $f = \sum_{I \in \mathcal{I}} \lambda_I(f) \cdot h_I$ in L^2 and

$$\|f\|_2^2 = \sum_I |\lambda_I|^2. \tag{1}$$

We define the *dyadic square function* of f by:

$$S(f)(x) = \left(\sum_{I \ni x} \frac{|\lambda_I(f)|^2}{|I|} \right)^{1/2}.$$

(As usual, $|\cdot|$ denotes Lebesgue measure.) Clearly, if $f \in L^2$, then $\|f\|_2 = \|S(f)\|_2$. (It is also well known (but harder to prove) that $\|f\|_p \sim \|S(f)\|_p$ for $1 < p < \infty$.)

In this talk we shall be concerned with inequalities that hold between certain *weighted* L^2 norms of f and $S(f)$. Such inequalities were used by Fefferman and Phong in their study of the Schrödinger operators $-\Delta - V$. Before we say any more about this, it is probably best to give an example.

---

* Partially supported under NSF Grant DMS-8811775.

A well-known weighted norm inequality for the square function is the following: Let $r > 1$ and let M be the usualy Hardy-Littlewood maximal operator. There is a constant C_r such that for all $f \in C_0^\infty(\mathbf{R})$ and all non-negative locally integrable V,

$$\int |f|^2 V \, dx \le C_r \int S^2(f) \, (M(V^r))^{1/r} \, dx. \tag{2}$$

To save eyestrain, we will henceforth denote $(M(V^r))^{1/r}$ by $M_r V$.

The reason that (2) holds is that $M_r V$ belongs to the Muckenhoupt A_∞ class, and therefore (2) holds even with $M_r V \ge V$ on the left-hand side.

Inequality (2) has an immediate corollary, which looks like (and is closely related to) a weighted Sobolev inequality. Denote the length of I by $\ell(I)$: Let $r > 1$ and $\alpha > 0$. If V is a non-negative function such that

$$\left(\frac{1}{|I|} \int_I V^r \, dx \right)^{1/r} \le \ell(I)^{-\alpha} \tag{3}$$

for all dyadic intervals I, then

$$\int |f|^2 V \, dx \le C(\alpha, r) \sum_I |\lambda_I(f)|^2 \ell(I)^{-\alpha} \tag{4}$$

for all $f \in C_0^\infty(\mathbf{R})$.

To see how (2) and (3) imply (4), we first rewrite (2) to get:

$$\int |f|^2 V \, dx \le C_r \sum_I |\lambda_I(f)|^2 \frac{1}{|I|} \int_I M_r V \, dx. \tag{5}$$

Thus, (4) will follow if we show that (3) implies

$$\frac{1}{|I|} \int_I M_r V \, dx \le C(\alpha, r) \ell(I)^{-\alpha} \tag{6}$$

for all I. This is very easy and very standard, but we will sketch the argument here, because it is going to be important later.

Let $\tilde{I}$ be the triple of I, and write $V = V_1 + V_2$, where $V_1 = V \cdot \chi_{\tilde{I}}$. Because of (3), $M_r V_2$ satisfies a (trivial) pointwise estimate $M_r V_2 \le C\ell(I)^{-\alpha}$ on I. The weak-type inequality for the maximal operator implies that

$$\frac{1}{|I|} \int_I M_r V_1 \le C_r \left(\frac{1}{|\tilde{I}|} \int_{\tilde{I}} V_1^r \, dx \right)^{1/r}.$$

Putting the two together yields (6), and hence (4).

In [F], Fefferman showed how (4) could be used to get a weighted Sobolev inequality. For $\beta > 0$, define the operator D^β:

$$\widehat{D^\beta f}(\xi) = |\xi|^\beta \hat{f}(\xi).$$

Fefferman showed that, if V satisfies (3), then

$$\int |f|^2 V \, dx \leq C(\alpha, r) \int |D^{\alpha/2} f|^2 \, dx \tag{7}$$

for all $f \in C_0^\infty(\mathbf{R})$. This is so because, with a little work, an object very much like the right-hand side of (4) can be shown to be dominated by the right-hand side of (7) (see [F]).

If we take $\alpha = 2$, then we get the Fefferman-Phong result on the positivity of the Schrödinger operator $L = -\Delta - V^1$: *Let $r > 1$. There is a positive constant $\gamma(r,d)$ such that if*

$$\ell(Q)^2 \left(\frac{1}{|Q|} \int_Q V^r \, dx \right)^{1/r} \leq \gamma(r,d)$$

for all cubes $Q \subset \mathbf{R}^d$, then $L \geq 0$; i. e.,

$$\int_{\mathbf{R}^d} |f|^2 V \, dx \leq \int |\nabla f|^2 \, dx$$

for all $f \in C_0^\infty(\mathbf{R}^d)^2$.

There is nothing that prevents us from applying this analysis to higher-order operators. Write $\mathbf{R}^d = \mathbf{R}^{d_1} \times \mathbf{R}^{d_2}$ and let Δ_i denote the Laplacian in the $\mathbf{R}^{d_i}$ variables. For $\delta > 0$, let us consider a "model" fourth-order operator $L_\delta = \delta(\Delta)^2 + \Delta_1\Delta_2 - V$. If δ is large then the first term dominates the cross term $\Delta_1\Delta_2$, and we can get information about L_δ's spectrum by applying the Fefferman-Phong methods to $\delta(\Delta)^2 - V$. But as $\delta \to 0$ the Fefferman-Phong estimates become less and less useful. They give us no way of telling, for instance, whether $\Delta_1\Delta_2$ might be "positive enough" to keep $L_\delta \geq 0$, even as δ approaches zero.

3. Two parameters.

There is an object which looks like it might help us solve this problem. Let $\mathcal{R}$ denote the family of rectangles in $\mathbf{R}^2$ which are Cartesian products of dyadic intervals. For each $R \in \mathcal{R}$, $R = I \times J$, define $\tilde{h}_R(x, y) = h_I(x) \cdot h_J(y)$, where h_I and h_J are the Haar functions defined in section 2. For each locally integrable f we analogously define $\tilde{h}_R(f) = \langle f, \tilde{h}_R \rangle$. The *two parameter square function* is given by:

$$\tilde{S}(f)(x, y) = \left(\sum_{R \ni (x,y)} \frac{|\tilde{\lambda}_R(f)|^2}{|R|} \right)^{1/2}.$$

As with the one-parameter square function, it is easy to see that $\|f\|_2 = \|\tilde{S}(f)\|_2$ if $f \in L^2$, and it is well known (but not so easy) that $\|f\|_p \sim \|\tilde{S}(f)\|_p$ for $1 < p < \infty$.

We seek a condition on V which will ensure that $L_0 = \Delta_1\Delta_2 - V \geq 0$. If we look at V's of the form $U(x_1) \cdot W(x_2)$, then the following conjecture seems reasonable: *Let*

---

[1] Recall that our applications are in higher dimensions!

[2] There is also a corresponding estimate for the bottom of L's spectrum.

$r > 1$. There is a $\gamma(r, d_1, d_2) > 0$ such that if, for every rectangle $R = Q_1 \times Q_2$, where the Q_i are cubes in $\mathbf{R}^{d_i}$, we have

$$\ell(Q_1)^2 \ell(Q_2)^2 \left(\frac{1}{|R|} \int_R V^r \, dx_1 \, dx_2 \right)^{1/r} \leq \gamma(r, d_1, d_2), \tag{8}$$

then $L_0 \geq 0$.

Let us try to prove this, using the method which worked in the one-parameter case.

If $q > 1$, then a fairly straightforward iteration of the one-parameter inequality yields:

$$\int_{\mathbf{R}^2} |f|^2 \, V \, dx_1 \, dx_2 \leq C_q \int_{\mathbf{R}^2} \tilde{S}^2(f) \, (M_1 M_2(V^q))^{1/q} \, dx_1 \, dx_2, \tag{9}$$

where M_i denotes the Hardy-Littlewood maximal operator taken with respect to the variable x_i. Henceforth we shall denote the maximal operator which occurs on the right-hand side of (9) as $\mathcal{M}_q V$. Rewriting (9), we get

$$\int_{\mathbf{R}^2} |f|^2 \, V \, dx_1 \, dx_2 \leq C_q \sum_{R \in \mathcal{R}} \frac{|\tilde{\lambda}_R(f)|^2}{|R|} \int_R \mathcal{M}_q V \, dx_1 \, dx_2.$$

The trick now is to show that (8) implies

$$\frac{1}{|R|} \int_R \mathcal{M}_q V \, dx_1 \, dx_2 \leq c\ell(Q_1)^{-2} \ell(Q_2)^{-2}. \tag{10}$$

As before, we let $\tilde{R}$ denote the triple of R, and write $V = V_1 + V_2$, with $V_1 = V\chi_{\tilde{R}}$. If we take $r > q$, then a simple iteration gives us the right bound on $\mathcal{M}_q V_1$. It is the V_2 term–which, recall, was trivial in the one-parameter case–that gives us trouble here. If $Q \subset \mathbf{R}^d$ is a cube, and Q^* is a second cube which touches both Q and $\mathbf{R}^d \setminus \tilde{Q}$, then $\ell(Q^*) \geq \ell(Q)$; but no such relation exists for rectangles $R \subset \mathbf{R}^{d_1} \times \mathbf{R}^{d_2}$ and the products of their sidelengths $\ell(Q_1)\ell(Q_2)$. In other words, this naive extension of the Fefferman-Phong method breaks down here.

If we do a little "proof-genealogy," we see that the source of our problem is in inequality (5). It is because $\int_Q M_r V \, dx$ reflects V's behavior off Q that we are forced to try to control $\mathcal{M}_q V_2$ in (10). We might have avoided this difficulty if we had been able to replace $\int_Q M_r V \, dx$ by the much cleaner-looking $\left(\frac{1}{|Q|} \int_Q V^r \, dx \right)^{1/r}$ in (5).

It turns out that this can be done. In fact, we can do much better.

Theorem 1. Let V_I denote the average of V over the interval I. Let $\epsilon > 0$. There is a C_ϵ such that for all $f \in C_0^\infty(\mathbf{R})$ and all weights V,

$$\int |f|^2 \, V \, dx \leq C_\epsilon \sum_{I \in \mathcal{I}} |\lambda_I(f)|^2 \cdot \frac{1}{|I|} \int_I V(x) \log^{1+\epsilon}(e + V(x)/V_I) \, dx.$$

We will sketch the proof; details can be found in [W1].

The proof begins with a remarkable theorem due to H. Rubin and T. Wolff [CWW]:
There are positive constants α and β such that if $\|S(f)\|_\infty \leq 1$ then

$$\frac{1}{|I|} \int_I \exp(\alpha|f - f_I|^2)\, dx \leq \beta \tag{11}$$

for all dyadic intervals I.

Now, (11) has an immediate consequence for weighted norm inequalities. Let us set

$$Y(I,V) = \begin{cases} \int_I V(x) \log(e + V(x)/V_I)\, dx / \int_I V\, dx & \text{when } \int_I V\, dx \neq 0; \\ 1 & \text{otherwise.} \end{cases}$$

By carefully using (11) in a Burkholder-Gundy "good-λ inequality," we get the following:
There is an absolute constant C such that if $Y(I,V) \leq A$ for all I then

$$\int |f|^2 V\, dx \leq CA \int S^2(f) V\, dx \tag{12}$$

for all f that have reasonable decay at infinity.

It is not hard to see that if $Y(I,V) \leq A$ for all I such that $\lambda_I(f) \neq 0$, then (12) follows, with the same C (see [W1]).

The proof of Theorem 1 is easy now. We may assume that V is bounded. For $k = 0, 1, 2, \ldots$, let $\mathcal{F}_k = \{I \in \mathcal{I}: 2^k \leq I < 2^{k+1}\}$, and set $f_k = \sum_{I \in \mathcal{F}_k} \lambda_I(f) \cdot h_I(x)$. The Cauchy-Schwarz inequality and the sharpened form of (12) imply that

$$\int |f|^2 V\, dx \leq C_\epsilon \sum_k 2^{k\epsilon} \int |f_k|^2 V\, dx$$

$$\leq C_\epsilon \sum_k 2^{(1+\epsilon)k} \int S^2(f_k) V\, dx$$

$$\leq C_\epsilon \sum_{I \in \mathcal{I}} |\lambda_I(f)|^2 \cdot Y(I,V)^{1+\epsilon} \int_I V\, dx$$

$$\leq C_\epsilon \sum_{I \in \mathcal{I}} |\lambda_I(f)|^2 \frac{1}{|I|} \int_I V(x) \log^{1+\epsilon}(e + V(x)/V_I)\, dx,$$

where the last line is from Hölder's inequality.

Now, using Theorem 1, we can obtain the following result for the two-parameter square function (see [W3], Theorem 2.1).

Theorem 2. *Let V_R denote V's average over the rectangle R. Let $\epsilon > 0$. There is a constant C_ϵ such that for all $f \in C_0^\infty(\mathbf{R}^2)$ and all weights V,*

$$\int |f|^2 V\, dx_1\, dx_2 \leq C_\epsilon \sum_{R \in \mathcal{R}} |\tilde{\lambda}_R(f)|^2 \cdot \frac{1}{|R|} \int_R V(x_1, x_2) \log^{2+\epsilon}(e + V(x_1, x_2)/V_R)\, dx_1,\, dx_2.$$

After some (technical and fairly tedious) work, we obtain our positivity result for L_0 (Theorem 4.1, [W3]):

Theorem 3. *Let* $\mathbf{R}^d = \mathbf{R}^{d_1} \times \mathbf{R}^{d_2}$ *and let* $\epsilon > 0$. *There is a positive constant* $C(\epsilon, d_1, d_2)$ *such that if,* V *is a weight and*

$$\frac{1}{|R|} \int_R V(x_1, x_2) \log^{2+\epsilon}(e + V(x_1, x_2)/V_R)\, dx_1\, dx_2 \leq C(\epsilon, d_1, d_2)\ell(Q_1)^{-2}\ell(Q_2)^{-2}$$

for all rectangles $R = Q_1 \times Q_2 \subset \mathbf{R}^d$, *then* $L_0 \geq 0$.

References.

[CWW] S. Y. A. Chang, J. M. Wilson, T. H. Wolff, "Some weighted norm inequalities concerning the Schrödinger operators," *Comm. Math. Helv.* **60** (1985), 217-246.

[F] C. L. Fefferman, "The uncertainty principle," *Bull. Amer. Math. Soc.* (N.S.) **9** (1983), 129-206.

[W1] J. M. Wilson, "A sharp inequality for the square function," *Duke Math. Journal* **55** (1987), 879-887.

[W2] —, "Weighted norm inequalities for the continuous square function," *Trans. Amer. Math. Soc.* **314** (1989), 661-692.

[W3] —, "Some two-parameter square function inequalities," to appear in *Indiana University Math. Journal*.

Two-Weight Norm Inequalities for the Fourier Transform

J. Michael Wilson*

Department of Mathematics

University of Vermont

Burlington, Vermont 05405

In this short talk we shall give two (trivial) necessary and two similar-looking sufficient conditions on non-negative weights V and W for the inequality

$$\int_{\mathbf{R}^d} |f(x)|^2\, V\, dx \le \int_{\mathbf{R}^d} |\hat{f}(\xi)|^2\, W\, d\xi \tag{1}$$

to hold for all $f \in C_0^\infty(\mathbf{R}^d)$.

We shall first take $d = 1$. Let ϵ be positive and small. For any integer k, let $I_k = \{\xi \in \mathbf{R}\colon 2^{k-1}(1-\epsilon) \le |\xi| \le 2^k(1+\epsilon)\}$. By looking at translates and dilates of a bump function, we see that for (1) to hold we must have, for each k,

$$\frac{1}{|I|}\int_I V(x)\,dx \le C \frac{1}{|I_k|}\int_{I_k} W(\xi)\,d\xi, \tag{2}$$

for all intervals I such that $\ell(I) = 2^{-k}$, with a constant C that does not depend on k or I.

Our first theorem says that the necessary condition (2)–with the addition of an extra hypothesis on W–is close to being sufficient.

Let us say that W is *weakly A_1* if there is a positive constant B such that

$$\frac{1}{|I_k|}\int_{I_k} W\,d\xi \le B \inf_{I_k} W(\xi)$$

for all integers k.

Theorem 1. *Let V and W be weights, and let $\delta > 0$. Suppose that W is weakly A_1, with corresponding constant B. There is a positive constant $C(\delta, B)$ such that if*

$$\frac{1}{|I|}\int_I V(x)\log^{1+\delta}(e + V(x)/V_I)\,dx \le C(\delta,B)\frac{1}{|I_k|}\int_{I_k} W(\xi)\,d\xi \tag{3}$$

holds for all k and all dyadic intervals I with $\ell(I) = 2^{-k}$, then (1) holds for all $f \in C_0^\infty(\mathbf{R}^d)$.

Proof. Let ψ be a non-trivial Schwartz function with $\operatorname{supp}\hat{\psi} \subset \{\xi\colon 1-\epsilon \le |\xi| \le 1+\epsilon\}$. Let η be a Schwartz function satisfying $\int \eta = 0$ and with support contained in $\{x\colon |x| \le 1\}$. The functions ψ and η can be chosen so that

$$\int_0^\infty \hat{\psi}(t\xi)\cdot\hat{\eta}(t\xi)\,\frac{dt}{t} = 1 \tag{4}$$

* Partially supported under NSF grant DMS-8811775.

for all $\xi \neq 0$. Let us assume that ψ and η have been so chosen. (This is the famous "Calderón-Torchinsky recipe." See, e. g., [Ca].) If $f \in C_0^\infty$ then (4) implies

$$f(x) = \int_{\mathbf{R}_+^2} (f * \psi_y(t)) \cdot \eta_y(x - t) \frac{dt\, dy}{y},$$

where ψ_y and η_y denote the usual L^1 dilates.

For every dyadic interval $I \subset \mathbf{R}$, let $T(I) = \{(x, y): x \in I, \frac{\ell(I)}{2} < y \le \ell(I)\}$, and set

$$b_I(x) = \int_{T(I)} (f * \psi_y(t)) \cdot \eta_y(x - t) \frac{dt\, dy}{y}.$$

We can write $b_I(x) = \lambda_I \cdot a_I(x)$, where λ_I is a constant and $a_I(x)$ is a function such that:

    $i)$ $\mathrm{supp}\, a_I \subset \tilde{I}$, the triple of I;
    $ii)$ $\int a_I = 0$;
    $iii)$ a_I is C^1, with $\|a_I'\|_\infty \le \ell(I)^{-1} \cdot |I|^{1/2}$.

Furthermore, we can take λ_I to satisfy

$$|\lambda_I| \le C \left(\int_{T(I)} |f * \psi_y(t)|^2 \frac{dt\, dy}{y} \right)^{1/2}.$$

Now we must use a result from [W1] and [W2]: if $f \in C_0^\infty$ has a decomposition $f = \sum \lambda_I \cdot a_I(x)$, where the a_I have the properties just mentioned, then, for any $\delta > 0$,

$$\int |f|^2 V\, dx \le C_\delta \sum_I |\lambda_I|^2 \cdot \frac{1}{|\tilde{I}|} \int_{\tilde{I}} V(x) \log^{1+\delta}(e + V(x)/V_{\tilde{I}})\, dx. \tag{5}$$

Rewrite the right-hand side of (5) as

$$\sum_{k=-\infty}^{\infty} \sum_{\ell(I)=2^{-k}} |\lambda_I|^2 \cdot \frac{1}{|\tilde{I}|} \int_{\tilde{I}} V(x) \log^{1+\delta}(e + V(x)/V_{\tilde{I}})\, dx \tag{6}$$

and temporarily fix k. If (4) holds for the intervals I, then it will also hold (with a slightly larger constant) for the intervals $\tilde{I}$. Thus the kth summand in (6) is dominated by

$$C \cdot \left(\sum_{\ell(I)=2^{-k}} |\lambda_I|^2 \right) \cdot \inf_{I_k} W(\xi). \tag{7}$$

By Plancherel's Theorem, the central sum in (7) is equal to

$$\int |\hat{f}(\xi)|^2 \left(\int_{2^{-k-1} < t \le 2^{-k}} |\hat{\psi}(t\xi)|^2 \frac{dt}{t} \right) d\xi,$$

which, because of the support restriction on $\hat{\psi}$, is dominated by

$$C \int_{I_k} |\hat{f}(\xi)|^2 \, d\xi.$$

The theorem follows after we plug this into (7) and sum on k. QED.

This result has a trivial generalization to $\mathbf{R}^d$, with the intervals I replaced by cubes and the I_k replaced by annuli; we leave the details of this to the interested reader.

The proof of Theorem 1 depended on a certain weighted norm inequality for the one-parameter square function. Because of the results in [W2], we know that analogous results also hold for the two-parameter square function. These immediately give us a *non*-trivial generalization of Theorem 1 to $\mathbf{R} \times \mathbf{R}$ (or, in general, $\mathbf{R}^{d_1} \times \mathbf{R}^{d_2}$), in which the integral and A_1 conditions on intervals are replaced by corresponding conditions on rectangles. For the sake of simplicity we shall state our result in $\mathbf{R} \times \mathbf{R}$.

For each ordered pair of integers (k,j), let $R_{k,j} = \{(\xi_1, \xi_2) \in \mathbf{R}^2 \colon \xi_1 \in I_k, \, \xi_2 \in I_j\}$, where I_k and I_j are as defined above. We shall say that W is *weakly* $A_1^{(2)}$ if there is a positive B such that

$$\frac{1}{|R_{k,j}|} \int_{R_{k,j}} W \, d\xi \le B \inf_{R_{k,j}} W(\xi)$$

for all (k,j).

By a *double-dyadic rectangle*, we shall mean a Cartesian product $R = I \times J$, where I and J are dyadic intervals in $\mathbf{R}$.

Theorem 2. *Let W and V be weights on $\mathbf{R}^2$, and let $\delta > 0$. Suppose that W is weakly $A_1^{(2)}$ with constant B. There is a positive constant $C(\delta, B)$ such that if*

$$\frac{1}{|R|} \int_R V(x,y) \log^{2+\delta}(e + V(x,y)/V_R) \, dx \, dy \le C(\delta, B) \frac{1}{|R_{k,j}|} \int_{R_{k,j}} W(\xi_1, \xi_2) \, d\xi_1 \, d\xi_2$$

holds for all double-dyadic rectangles $R = I \times J$, with $\ell(I) = 2^{-k}$, $\ell(J) = 2^{-j}$, for all k and j, then

$$\int_{\mathbf{R}^2} |f|^2 \, V \, dx \, dy \le \int_{\mathbf{R}^2} |\hat{f}|^2 \, W \, d\xi_1 \, d\xi_2 \tag{8}$$

holds for all $f \in \mathcal{C}_0^\infty(\mathbf{R}^2)$.

Remark: We note that, by considering tensor products of bump functions, a trivial necessary condition for (8) is that

$$\frac{1}{|R|} \int_R V \, dx \, dy \le \frac{c}{|R_{k,j}|} \int_{R_{k,j}} W \, d\xi_1 \, d\xi_2$$

hold for all double-dyadic rectangles $R = I \times J$ with $\ell(I) = 2^{-k}$ and $\ell(J) = 2^{-j}$, and all k and j.

Proof. By iterating the Calderón-Torchinsky procedure in x and y, we obtain a decomposition of our function f:

$$f = \sum_R \lambda_R \cdot a_R(x,y), \tag{9}$$

where the sum is over all double-dyadic rectangles $R = I \times J$, and the λ_R's are constants satisfying

$$|\lambda_R| \le C \left(\int_{T(I) \times T(J)} |f * \psi_{y_1}(t_1) * \psi_{y_2}(t_2)|^2 \frac{dt_1 \, dt_2 \, dy_1 \, dy_2}{y_1 \, y_2} \right)^{1/2}.$$

The functions $a_R(x, y)$ are two-parameter analogues of the a_I's in the proof of Theorem 1. They satisfy the following three conditions:

i) $\mathrm{supp}\, a_R \subset \tilde{R} = \tilde{I} \times \tilde{J}$;

ii) for each fixed x, $\int a_R(x, y)\, dy = 0$, and for each fixed y, $\int a_R(x, y)\, dx = 0$;

iii) $a_R(x, y)$ is C^2, and $\|\frac{\partial^2 a_R}{\partial x \partial y}\|_\infty \le \ell(I)^{-1} \ell(J)^{-1} |R|^{-1/2}$.

In [W2] it is proved that if $f \in \mathcal{C}_0^\infty(\mathbf{R}^2)$, and has a decomposition (9) with the foregoing properties, then for any weight V and any $\delta > 0$,

$$\int |f|^2 \, dx \, dy \le C_\delta \sum_R |\lambda_R|^2 \frac{1}{|\tilde{R}|} \int_{\tilde{R}} V(x, y) \log^{2+\delta}(e + V(x, y)/V_{\tilde{R}}) \, dx \, dy.$$

Theorem 2 now follows, with almost the same proof as Theorem 1.

References.

[Ca] A. P. Calderón, "An atomic decomposition of distributions in parabolic H^p spaces," *Advances in Mathematics* **25** (1977), 216-225.

[W1] J. M. Wilson, "Weighted norm inequalities for the continuous square function," *Trans. Amer. Math. Soc.* **314** (1989), 661-692.

[W2] ——, "Some two-parameter square function inequalities," to appear in *Indiana University Math. Journal.*

PROGRAM

14 AUGUST
11:00-11:50   P. JONES (Yale Univ.), The travelling salesman problem and related problems in harmonic analysis
13:30-14:20   P. MATTILA (Univ. of Jyväskylä), Singular Cauchy integrals and rectifiability of measures in the plane
14:30-15:20   P. G. LEMARIÉ (Univ. de Paris-Sud), Wavelet basis on nilpotent Lie groups
16:00-16:50   A. L. VOL'ERG (Steklov Inst.), On a generalized Fourier transform and its use in convolution equation

15 AUGUST
9:30 -10:20   J.-P. KAHANE (Univ. de Paris-Sud), From Riesz products to random sets
11:00-11:50   S.-Y. A. CHANG (UCLA), Extremal inequalities for log determinant of conformal Laplacian on S^n
13:30-14:20   J. M. WILSON (Univ. of Vermont), Littlewood-Paley theory in one and two parameters
14:30-15:20   A. MIYACHI (Hitotsubashi Univ.), Maximal functions and function spaces on Euclidean domains
16:00-16:50   J. M. ASH (De Paul Univ.), Various proofs of uniqueness of representation by trigonometric series

16 AUGUST
9:30 -10:20   H. ARAI (Tohoku Univ.), Harmonic analysis with respect to degenerate Laplacians on strictly pseudoconvex domains
11:00-11:50   M. CHRIST (UCLA), Remarks on $\overline{\partial}$ and $\overline{\partial}_b$
13:30-14:20   F. SORIA (Univ. Autónoma), Weights and vector valued inequalities for the disk multiplier
14:30-15:20   G. GAUDRY (Flinders Univ.), Some singular integrals on the affine groups
16:00-16:50   J.-P. ANKER (Cornell Univ.), Sharp estimates for some functions of the Laplacian on non-compact symmetric spaces

17 AUGUST
9:30 -10:20   W. RUDIN (Univ. of Wisconsin), M-harmonic products
11:00-11:50   R. R. COIFMAN (Yale Univ.), L^2 estimates in nonlinear Fourier analysis

13:30-15:20   SHORT TALKS:
R. ASKEY (Univ. of Wisconsin), The orthogonal polynomials associated with quantum group $SU_q(2)$
G. ALEXOPOULOS (McGill Univ.), Harmonic analysis on Lie groups of polynomial
T. KAWAZOE (Keio Univ.), A method of reduction in harmonic analysis on real rank 1 semisimple Lie groups
J. WILSON (Univ. of Vermont), Some 2-weight norm inequalities for the Fourier transform

T. QIAN (Flinders Univ.), Clifford martingale Φ-inequality, Clifford $T(b)$ theorem and singular integrals with monogenic kernels on Lipshitz surfaces

S.-Z. LU (Beijing Normal Univ.), Oscillatory singular integrals with rough kernel

V. BURENKOV (Friendship of Nations Univ.), Fourier multipliers in spaces with exponential weights and their applications to hypoelliptic equations

L. HEDBERG (Univ. of Linköping), Compactness of intervals in some spaces of distributions

J. GARCIA-CUERVA (Univ. Autónoma), Extrapolation theory : endpoint results and applications

J. RNO (Univ. of Cincinnati), Harmonic analysis on the Euclidean group in three space

S. EMARA (Amer.Univ. in Cairo), A class of weighted inequalities

DINH-DUNG (Inst. Computer Sci.), On harmonic approximation of multivariate functions

16:00-16:50   SHORT TALKS:

C.-M. LEE (Univ. of Wisconsin-Milwaukee), A symmetric approximate Perron integral for the coefficient problem of trigonometric series

N. SHIEH (National Taiwan Univ.), Self-intersections of Markov processes

N. SHIMAKURA (Tohoku Univ.), Elementary solution for Bessel operator on matrix space

T. MIZUHARA (Yamagata Univ.), Generalized Morrey spaces and Hardy-Littlewood maximal functions

G. SINNAMON (Univ.of Western Ontario), Spaces defined by the level function and their duals

J.N. PANDEY (Carleton Univ.),· Characterization of functions with Fourier transform supported on orthants

18 AUGUST

9:30 -10:20   N. Th. VAROPOULOS (Univ. de Paris IV), Subelliptic distances and pseudodifferential operators

11:00-11:50   L. VEGA (Univ. of Chicago), Oscillatory integrals and the Korteweg-de Vries equation

LIST OF PARTICIPANTS

AKIBA, Yûitirô (Japan)

ALEXOPOULOS, George (McGill U., Canada)

ANKER, Jean-Philippe (Cornell U., USA)

ANZAI, Kazuo (Kagawa U., Japan)

ARAI, Hitoshi (Tohoku U., Japan)

ARAI, Takahiro (Keio U., Japan)

ASH, J. Marshall (De Paul U., USA)

ASKEY, Richard A. (U. of Wisconsin, USA)

ATSUJI, A. (Tokyo U., Japan)

AZUMA, Kazuoki (Miyagi U. of Educ., Japan)

BURENKOV, Victor (FNU, USSR)

CHANG, Sun-Yung A. (UCLA, USA)

CHEE, P. S. (U. Malaya, Malaysia)

CHRIST, F. Michael (UCLA, USA)

COIFMAN, Ronald R. (Yale U., USA)

DINH-DUNG (Inst. of Computer Sci., Vietnam)

DUOANDIKOETXEA, Javier (U. Autónoma, Spain)

EMARA, Salah A. A. (The American U., Egypt)

ESSÉN, Matts R. (U. of Uppsala, Sweden)

FUJII, Nobuhiko (Tokai U., Japan)

FUKUYAMA, K. (U. of Tsukuba, Japan)

GARCIA-CUERVA, José (U. Autónoma, Spain)

GAUDRY, Garth I. (Flinders U., Australia)

GELLER, D. (SUNY at Stony Brook, USA)

GOTOH, Yasuhiro (Kyoto U., Japan)

HAHN, Liang-Shin (U. of New Mexico, USA)

HASEGAWA, Yoshimitsu (Hirosaki U., Japan)

HASHIMOTO, Toshio (Hokkaido U., Japan)

HATORI, Osamu (Tokyo Medical Coll., Japan)

HAYASHI, Mikihiro (Hokkaido U., Japan)

HEDBERG, Lars I. (U. of Linköping, Sweden)

IGARI, Satoru (Tohoku U., Japan)

IKENO, Kyuhei (Hachinohe U., Japan)

INOUE, Jyunji (Hokkaido U., Japan)

ISHIKAWA, Shiro (Keio U., Japan)

ITO, Yasuyuki (Hokkaido U., Japan)

IZUCHI, Keiji (Kanagawa U., Japan)

IZUMISAWA, Masataka (Tottori U., Japan)

JONES, Peter W. (Yale U., USA)

KAHANE, Jean-Pierre (U. de Paris-Sud, France)

KANEKO, Makoto (Tohoku U., Japan)

KANJIN, Yuichi (Kanazawa U., Japan)

KATSUMATA, Osamu (Hokkaido U., Japan)

KAWAZOE, Takeshi (Keio U., Japan)

KAZAMAKI, Norihito (Toyama U., Japan)

KITA, Hiroo (Oita U., Japan)

KITADA, Toshiyuki (Hirosaki U., Japan)

KOBAYASHI, Atsushi (Waseda U., Japan)

KOIZUMI, Sumiyuki (Keio U., Japan)

KOJIMA, Michitaka (Hokuriku U., Japan)

KOMORI, Yasuo (Inagi Highschl., Japan)

KÔNO, Norio (Kyoto U., Japan)

KOSHI, Shozo (Hokkaido U., Japan)

KURIBAYASHI, Yukio (Tottori U., Japan)

KURODA, Shige Toshi (Gakushuin U., Japan)

KURODA, T. (Tohoku U., Japan)

LEE, Cheng-Ming (U. of Wisconsin-Milwaukee, USA)

LEMARIÉ, Pierre (U. de Paris-Sud, France)

LU, Shan-zhen (Beijing Normal U., China)

MATTILA, Pertti (U. of Jyväskylä, Finland)

MATSUBARA, Minoru (Tokyo Woman's Chris. U., Japan)

MATSUOKA, Katsuo (Keio Shiki High Schl., Japan)

MIYACHI, Akihiko (Hitotsubashi U., Japan)

MIZUHARA, Takahiro (Yamagata U., Japan)

MURAI, Takafumi (Nagoya U., Japan)

NAKAI, Eiichi (Yuki Diichi Senior High Schl., Japan)

NAKAMURA, Akihiro (Tokai U., Japan)

NAKAMURA, Shigeaki (Keio U., Japan)

NISHIGAKI, Sei-ichi (Numazu Coll. Tech., Japan)

NISHISHIRAHO, Toshihiko (U. Ryukyus, Japan)

OGIWARA, Etsuo (Hokkaido U., Japan)

OHNO, Yoshiki (Tohoku U., Japan)

OKADA, M. (Tohoku U., Japan)

OKADA, Tatsuya (Fukushima Med. Coll., Japan)

OKUYAMA, Yasuo (Shinshu U., Japan)

ONOE, Yoshikazu (Keio U., Japan)

OYA, Fumimasa (Tokai U., Japan)

PANDEY, Jagdish Narayan (Carleton U., Canada)

QIAN, Tao (Flinders U., Australia)

RNO, Jung S. (U. of Cincinnati, USA)

RUDIN, Mary E. (USA)

RUDIN, Walter (U. of Wisconsin, USA)

SAITO, Yasuhiko (Tohoku U., Japan)

SAKA, Kōichi (Akita U., Japan)

SATAKE, Makoto (Kanazawa Women's Junior Coll. Japan)

SATO, Enji (Yamagata U., Japan)

SATO, Hiroshi (Kyushu U., Japan)

SATO, Kunio (Yamagata U., Japan)

SATO, Noriyuki (Tohoku U., Japan)

SATO, Shuichi (Kanazawa U., Japan)

SATO, Tokushi (Tohoku U., Japan)

SHIEH, Narn R. (National Taiwan U., China)

SHIMAKURA, Norio (Tohoku U., Japan)

SHIMOMURA, Katsunori (Ibaraki U., Japan)

SHIOTA, Yasunobu (Akita U., Japan)

SINNAMON, Gordon J. (U. of Western Ontario, Canada)

SJÖGREN, Peter (Chalmers U., Sweden)

SJÖLIN, Per (Uppsala U., Sweden)

SOBUKAWA, Takuya (Keio U., Japan)

SORIA, Fernando (U. Autónoma, Spain)

SUNOUCHI, Gen-ichiro (Japan)

SUZUKI, Yoshiya (Tohoku U., Japan)

TACHIZAWA, Kazuya (Tohoku U., Japan)

TAKAGI, Hiroyuki (Waseda U., Japan)

TAKAGI, Izumi (Tohoku U., Japan)

TAKAYAMA, Takuma (Waseda U., Japan)

TAKEUCHI, Yukio (Tohoku U., Japan)

TANAKA, Jun-ichi (Tsuru U., Japan)

TATEOKA, Jun (Akita U., Japan)

TORRES, Rodolfo H. (Washington U., USA)

TSUCHIKURA, T. (Tokyo Denki U., Japan)

VAROPOULOS, Th. N. (U. de Paris VI, France)

VEGA, Luis (U. of Chicago, USA)

VOL'BERG, Alexander L. (Steklov Math. Inst., USSR)

WADA, Junzo (Waseda U., Japan)

WATANABE, Hiroshi (Tokai U., Japan)

WATARI, Chinami (Tohoku Gakuin U., Japan)

WILSON, James M. (U. of Vermont, USA)

YABUTA, Kozo (Nara Women's U., Japan)

YAMAGUCHI, Hiroshi (Josai U., Japan)

YAMAMOTO, Haruyuki (Tohoku U., Japan)

YAMAZAKI, Masao (Hitotsubashi U., Japan)

YANO, Shigeki (Japan)

YOKOI, Hirofumi (Tohoku U., Japan)

YONEDA, Kaoru (Osaka Prefectural U., Japan)

YOSHIDA, Kaori (Tohoku U., Japan)

YOSHIDA, Nobuo (Kyoto U., Japan)